# Fifth Workshop on Building and Evaluating Resources for Biomedical Text Mining (BioTxtM 2016)

Osaka, Japan
11 - 16 December 2016

ISBN: 978-1-5108-3331-9

# Preface

Biomedical natural language processing has grown from its roots in clinical language processing and bioinformatics into a thriving research field of its own. The search *("natural language processing") OR ("text mining")* performed in PubMed today returns 5,056 hits, versus 1,903 at the turn of the century and 3,485 in 2010. The papers appearing in this volume reflect the diversity of trends in biomedical natural language processing today—movement from English-language texts to clinical texts in other languages; exploration of social media in addition to clinical documents and traditional scientific publications; and processing of full-text articles, versus abstracts only. In addition to reflecting the diversity of the field, the papers in this volume also reflect the homogenisation of approaches that has characterised some recent approaches, with 4 out of 15 papers involving some combination of neural networks and/or distributional semantics. The organisers thank the authors for sharing their science with this community, and the programme committee (listed elsewhere in this volume) for their contribution to maintaining the high standards of the BioTxtM series of meetings.

## Keynote Talk by Dr. Makoto Miwa

*Learning for Information Extraction in Biomedical and General Domains*

Information extraction (IE) has been widely studied in various domains since IE is a key to bridge the gap between knowledge and texts. IE includes several core sub-problems, such as named entity recognition, relation extraction, and event extraction, and these sub-problems have been tackled using machine learning techniques. In this talk, I will give an overview of learning approaches for IE in biomedical and general domain, especially on corpus-based classification and structured learning approaches. I will then introduce recent deep learning approaches including our recent recurrent neural network (RNN)-based approach, and discuss the limitations and future directions.

## Speaker Biography

Makoto Miwa is an associate professor of Toyota Technological Institute (TTI). He received his Ph.D. from the University of Tokyo in 2008. His research mainly focuses on information extraction from texts, deep learning, and representation learning. His projects include AkaneRE, EventMine, PathText and LSTM-ER.

**Organisers**

Sophia Ananiadou, National Centre for Text Mining, University of Manchester UK

Riza Batista-Navarro, National Centre for Text Mining, University of Manchester UK

Kevin Bretonnel Cohen, Computational Bioscience Program, University of Colorado School of Medicine, USA

Dina Demner-Fushman, National Library of Medicine, USA

Paul Thompson, National Centre for Text Mining, University of Manchester, UK

**Programme Committee**

Eiji Aramaki, Nara Institute of Science and Technology (NAIST), Japan

Hercules Dalianis, Stockholm University, Sweden

Graciela Gonzalez, Arizona State University, USA

Wen-Lian Hsu, Academia Sinica, Taipei, Taiwan

Rezarta Islamaj, NCBI/NLM/NIH, USA

Roman Klinger, University of Stuttgart, Germany

Robert Leaman NCBI/NLM/NIH, USA

Shervin Malmasi, Harvard Medical School, USA

Makoto Miwa, Toyota Technological Institute, Japan

Sung-Hyon Myaeng, Korea Advanced Institute of Science and Technology (KAIST), Korea

Claire Nedellec, French National Institute of Agronomy (INRA), France

Naoaki Okazaki, Tohoku University, Japan

Arzucan Özgür, Bogazici University, Turkey

Martha Palmer, University of Colorado at Boulder, USA

Stelios Piperidis, Institute for Language and Speech Processing, Greece

Guergana Savova, Boston Children's Hospital and Harvard Medical School, USA

Hagit Shatkay, University of Delaware, USA

Mark Stevenson, University of Sheffield, UK

Yoshimasa Tsuruoka, University of Tokyo, Japan

Lucy Vanderwende, Microsoft, USA

Karin Verspoor, University of Melbourne, Australia

Stephen Wu, Oregon Health & Science University, USA

Yan Xu, Microsoft Research Asia, China

Pierre Zweigenbaum, Laboratoire d'Informatique pour la Mécanique et les Sciences de l'Ingénieur (LIMSI), France

# Table of Contents

# Conference Program

**12th December 2016**

**9:00–9:10**   **Welcome remarks**

**9:10–10:20**   **Session 1**

9:10–9:30   *Cancer Hallmark Text Classification Using Convolutional Neural Networks*
Simon Baker, Anna Korhonen and Sampo Pyysalo

9:30–9:50   *Learning Orthographic Features in Bi-directional LSTM for Biomedical Named Entity Recognition*
Nut Limsopatham and Nigel Collier

9:50–10:00   *Building Content-driven Entity Networks for Scarce Scientific Literature using Content Information*
Reinald Kim Amplayo and Min Song

10:00–10:10   *Named Entity Recognition in Swedish Health Records with Character-Based Deep Bidirectional LSTMs*
Simon Almgren, Sean Pavlov and Olof Mogren

10:10–10:20   *Entity-Supported Summarization of Biomedical Abstracts*
Frederik Schulze and Mariana Neves

**10:20–10:50**   **Coffee break and Poster Session 1**

**12th December 2016 (continued)**

**10:50–12:00   Session 2**

10:50–11:10   *Fully unsupervised low-dimensional representation of adverse drug reaction events through distributional semantics*
Alicia Pérez, Arantza Casillas and Koldo Gojenola

11:10–12:00   *Keynote Talk: Learning for Information Extraction in Biomedical and General Domains*
Dr. Makoto Miwa

**12:00–14:00   Lunch break**

**14:00–15:20   Session 3**

14:00–14:20   *A Dataset for ICD-10 Coding of Death Certificates: Creation and Usage*
Thomas Lavergne, Aurelie Neveol, Aude Robert, Cyril Grouin, Grégoire Rey and Pierre Zweigenbaum

14:20–14:40   *A Corpus of Tables in Full-Text Biomedical Research Publications*
Tatyana Shmanina, Ingrid Zukerman, Ai Lee Cheam, Thomas Bochynek and Lawrence Cavedon

14:40–14:50   *Supervised classification of end-of-lines in clinical text with no manual annotation*
Pierre Zweigenbaum, Cyril Grouin and Thomas Lavergne

14:50–15:00   *BioDCA Identifier: A System for Automatic Identification of Discourse Connective and Arguments from Biomedical Text*
Sindhuja Gopalan and Sobha Lalitha Devi

15:00–15:10   *Data, tools and resources for mining social media drug chatter*
Abeed Sarker and Graciela Gonzalez

15:10–15:20   *Detection of Text Reuse in French Medical Corpora*
Eva D'hondt, Cyril Grouin, Aurelie Neveol, Efstathios Stamatatos and Pierre Zweigenbaum

**12th December 2016 (continued)**

**15:20–15:50**    **Coffee break and Poster Session 2**

**15:50–16:50**    **Session 4**

15:50–16:10    *Negation Detection in Clinical Reports Written in German*
Viviana Cotik, Roland Roller, Feiyu Xu, Hans Uszkoreit, Klemens Budde and Danilo Schmidt

16:10–16:30    *Scoring Disease-Medication Associations using Advanced NLP, Machine Learning, and Multiple Content Sources*
Bharath Dandala, Murthy Devarakonda, Mihaela Bornea and Christopher Nielson

16:30–16:50    *Author Name Disambiguation in MEDLINE Based on Journal Descriptors and Semantic Types*
Dina Vishnyakova, Raul Rodriguez-Esteban, Khan Ozol and Fabio Rinaldi

**16:50–17:00**    **Closing remarks**

# Cancer Hallmark Text Classification
# Using Convolutional Neural Networks


**Simon Baker** [1,2]     **Anna Korhonen** [2]     **Sampo Pyysalo** [2]

[1]Computer Laboratory, 15 JJ Thomson Avenue
[2] Language Technology Lab, DTAL
University of Cambridge, UK
`simon.baker@cl.cam.ac.uk, alk23@cam.ac.uk, sampo@pyysalo.net`


## Abstract


Methods based on deep learning approaches have recently achieved state-of-the-art performance in a range of machine learning tasks and are increasingly applied to natural language processing (NLP). Despite strong results in various established NLP tasks involving general domain texts, there is only limited work applying these models to biomedical NLP. In this paper, we consider a Convolutional Neural Network (CNN) approach to biomedical text classification. Evaluation using a recently introduced cancer domain dataset involving the categorization of documents according to the well-established hallmarks of cancer shows that a basic CNN model can achieve a level of performance competitive with a Support Vector Machine (SVM) trained using complex manually engineered features optimized to the task. We further show that simple modifications to the CNN hyperparameters, initialization, and training process allow the model to notably outperform the SVM, establishing a new state of the art result at this task. We make all of the resources and tools introduced in this study available under open licenses from `https://cambridgeltl.github.io/cancer-hallmark-cnn/`.


## 1 Introduction

A major goal of cancer research is to understand the biological mechanisms involved in tumorous growths starting in the body, being sustained, and turning malignant. Cancer is often described in the biomedical literature by its *hallmarks*; a set of interrelated biological properties and behaviors that enable cancer to thrive in the body. The hallmarks of cancer were first introduced in the seminal paper of Hanahan and Weinberg (2000), the most cited paper in the journal *Cell*. The paper introduces six hallmarks, which were then extended in a follow-up paper (Hanahan and Weinberg, 2011) by another four, forming the set of ten hallmarks that are known today. The current set of hallmarks distill our knowledge of the disease into a fixed set of alterations in cell physiology that affect malignant growth, such as self-sufficiency in growth signals, insensitivity to growth-inhibitors, evasion of programmed cell death, limitless replicative potential, sustained angiogenesis, and tissue invasion.

In the context of biomedical text mining, the original six hallmarks of cancer were used as an organizing principle in the BioNLP Shared Task 2013 Cancer Genetics task (Pyysalo et al., 2013b), which involved the extraction of events (biological processes) from cancer domain texts. The hallmarks have also inspired other information extraction efforts and the development of tools such as *OncoSearch* (Lee, 2014) and *OncoCL* (Doland, 2014). In recent work, Baker et al. (2016) introduced a corpus comprised of over 1,800 abstracts from biomedical publications annotated with the ten hallmarks of cancer. Baker et al. also proposed a machine learning based method for classifying text according to the hallmarks. The approach utilizes a conventional NLP pipeline that extracts a feature-rich representation that is used to train support vector machine (SVM) classifiers. The method achieves a respectable level of performance, identifying hallmarks with an average F-score of 77%, but with the cost of involving a lengthy and computationally demanding NLP pipeline.



1


*Proceedings of the Fifth Workshop on Building and Evaluating Resources for Biomedical Text Mining (BioTxtM 2016),*
pages 1–9, Osaka, Japan, December 12th 2016.

In this work, our focus is on studying biomedical text classification using machine learning methods that emphasize *feature learning* rather than manual feature engineering. We adopt the task setting and dataset of Baker et al. (2016), but instead of SVMs, we focus on convolutional neural networks (CNN). CNNs were first proposed for image processing (LeCun and Bengio, 1995) and have been recently shown to achieve state-of-the-art performance in a range of NLP tasks, in particular in text classification (Zhang et al., 2015; Severyn and Moschitti, 2015; Zhang and Wallace, 2015). While neural network-based methods in general and "deep" networks in particular are increasingly popular for general domain NLP, there has been comparatively little work applying this class of methods to biomedical text. One recent study applying a CNN model to biomedical text classification task was presented by (Limsopatham and Collier, 2016), who applied CNNs to the task of adverse drug reaction detection in social media messages (Ginn et al., 2014). In addition to the specific subdomain of the source texts and the novel categories represented by the hallmarks of cancer, one factor that sets apart the task here from this previous work is the length of the texts: instead of sentences or brief social media messages, our task involves the classification of publication abstracts typically consisting of hundreds of words.

## 2  Data

For training and evaluating our methods, we use the corpus of 1852 biomedical publication abstracts annotated for the hallmarks of cancer by Baker et al. (2016). Each abstract in the dataset may be labeled with zero or more of the ten hallmarks, i.e. the task is multi-label classification. The ten hallmarks are summarized below:

**Sustaining proliferative signaling**: Healthy cells require molecules that act as signals for them to grow and divide. Cancer cells, on the other hand, are able to grow without these external signals.
**Evading growth suppressors**: Cells have processes that halt growth and division. In cancer cells, these processes are altered so that they don't effectively prevent cell division.
**Resisting cell death**: Apoptosis is a mechanism by which cells are programmed to die in the event that they become damaged. Cancer cells are able to bypass these mechanisms.
**Enabling replicative immortality**: Non-cancer cells die after a certain number of divisions. Cancer cells, however, are capable of indefinite growth and division (immortality).
**Inducing angiogenesis**: Cancer cells are able to initiate angiogenesis, the process by which new blood vessels are formed, thus ensuring the supply of oxygen and other nutrients.
**Activating invasion & metastasis**: Cancer cells can break away from their site of origin to invade surrounding tissue and spread to distant body parts.
**Genome instability & mutation**: Cancer cells generally have severe chromosomal abnormalities, which worsen as the disease progresses.
**Tumor-promoting inflammation**: Inflammation affects the microenvironment surrounding tumors, contributing to the proliferation, survival and metastasis of cancer cells.
**Deregulating cellular energetics**: Most cancer cells use abnormal metabolic pathways to generate energy, e.g. exhibiting glucose fermentation even when enough oxygen is present to properly respire.
**Avoiding immune destruction**: Cancer cells are invisible to the immune system.

We divide the dataset into ten binary-labeled datasets (one per hallmark), where the positive examples in each are the abstracts annotated with the hallmark, and negative examples are those that are not. While we generally aim to follow the experimental setup of Baker et al., we chose to split the annotated data into training, development and test subsets instead of applying the cross-validation setup of the study introducing the dataset. Cross-validation setups using all available data fail to make a clear separation between data used for method development and blind data held out for final testing only, and should be avoided in studies involving experimentally driven model refinement (as we do here). Consequently, we initially split the corpus in 70/10/20% proportion to train, development and test sets with a random sampling strategy that aimed to roughly preserve the overall class distribution in each sample. The test set was held out during development and only used in the final experiments. Table 1 shows the distribution of positive and negative labels for each hallmark.

| | Train | | Devel | | Test | | Total | |
| Hallmark | pos | neg | pos | neg | pos | neg | pos | neg |
|---|---|---|---|---|---|---|---|---|
| Sustaining proliferative signaling | 328 | 975 | 43 | 140 | 91 | 275 | 462 | 1390 |
| Evading growth suppressors | 172 | 1131 | 22 | 161 | 46 | 320 | 240 | 1612 |
| Resisting cell death | 303 | 1000 | 42 | 141 | 84 | 282 | 429 | 1423 |
| Enabling replicative immortality | 81 | 1222 | 11 | 172 | 23 | 343 | 115 | 1737 |
| Inducing angiogenesis | 99 | 1204 | 13 | 170 | 31 | 335 | 143 | 1709 |
| Activating invasion and metastasis | 208 | 1095 | 29 | 154 | 54 | 312 | 291 | 1561 |
| Genomic instability and mutation | 227 | 1076 | 38 | 145 | 68 | 298 | 333 | 1519 |
| Tumor promoting inflammation | 169 | 1134 | 24 | 159 | 47 | 319 | 240 | 1612 |
| Cellular energetics | 74 | 1229 | 10 | 173 | 21 | 345 | 105 | 1747 |
| Avoiding immune destruction | 77 | 1226 | 10 | 173 | 21 | 345 | 108 | 1744 |

Table 1: Annotation statistics

## 3 Methods

We implement and evaluate two SVM-based methods and two CNN variants, described in the following. All of these machine learning methods are applied to the multi-label task by training ten binary classifiers, one for each hallmark label.

### 3.1 SVM with Bag of Words Features

We implement a simple classifier using only Bag of Words (BoW) features as a basic SVM baseline. In the BoW approach each document is represented by the set of words appearing in it, discarding word order and frequency information. For training the model, we use the linear kernel SVM implemented in the Scikit-learn (Pedregosa et al., 2011) toolkit. We fine-tune the regularization hyperparameter $c$ conventionally using evaluation on the development dataset with a search between $10^{-2}$ and $10^2$ on a log scale.

### 3.2 SVM with Rich Features

For our primary point of reference, we replicated the NLP pipeline and SVM model of Baker et al. (2016) for hallmark classification. This model uses a rich set of features derived from the application of several state-of-the-art systems for biomedical NLP, summarized briefly in the following.[1]

**Lemmatized bag of words** All non-stop words in the documents are lemmatized using *BioLemmatizer* (Liu et al., 2012) and included as features using a BoW-style representation.

**Noun bigrams** Compound nouns (without lemmatization) are combined to generate bigram features. Nouns pairs often represent specific, discriminative concepts such as "*gene mutation*".

**Grammatical relations triples** The *C&C Parser* with a biomedical domain model (Rimell and Clark, 2009) is used to parse the documents, and the *dobj* (direct object), *ncsubj* (non-clausal subject) and *iobj* (indirect object) relations, and their head and dependent words then represented as features.

**Verb classes** The hierarchical classification of 399 verbs of Sun and Korhonen (2009) is used to generate features for verbs, utilizing all three levels of abstraction by allocating three bits in the feature representation for each concrete class, i.e. one bit for each level of the verb class hierarchy.

**Named entities (NE)** The *ABNER* NER tool (Settles, 2005) is used to identify five named entity types that are particularly relevant to cancer research: proteins, DNA, RNA, cell lines and cell types. Features are then created pairing each entity type and its associated words.

**Medical subject headings (MeSH)** The MeSH headings assigned to the documents in the biomedical publication indexing process are included as features using a bag-of-headings representation.

**Chemical lists** Similarly to MeSH terms, many documents are indexed with chemical identifiers. These identifiers are used analogously to the MeSH terms to generate features.

---

[1] We refer to Baker et al. (2016) for the further details on this feature representation.

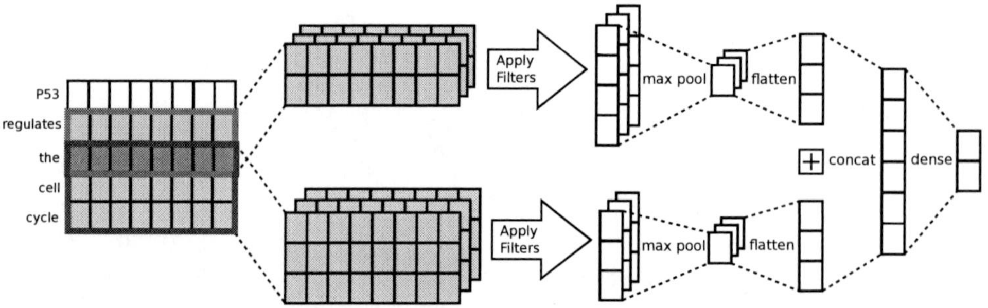


Figure 1: Network architecture

All features are extracted from the training data and are then filtered by frequency to remove features that are too common or too rare, leaving behind only the most discriminating features. We use a linear kernel and fine-tune the regularization parameter $c$ on the development dataset using the same process applied for the BoW model. As there are significantly more negatively labelled documents than positives, we use inverse class weighting in order to correct for the class imbalance when training the classifiers.

### 3.3 Convolutional Neural Network

We base our CNN architecture on the simple model of Kim (2014). In brief, this model consists of an initial embedding layer that maps input texts into matrices, followed by convolutions of different filter sizes and 1-max pooling, and finally a fully connected layer. The architecture is illustrated in Figure 1. We implemented the neural network using Keras (Chollet, 2015). Model hyperparameters and the training setup were initially based on those applied by Kim (2014), summarized in the following:

| Parameter | Value |
| --- | --- |
| Word vector size | 300 (Google News vectors) |
| Filter sizes | 3, 4, and 5 |
| Number of filters | 300 (100 of each size) |
| Dropout probability | 0.5 |
| Minibatch size | 50 |

Table 2: Kim (2014) model parameters

Some of these parameters were further refined in experiments using only the training and development portions of the data (see Section 4.1). In the final test set experiments, we evaluate the network using both the set of parameters used by Kim (2014) as well as with those selected in our development set experiments. We train the models for 20 epochs using categorical cross-entropy loss and the Adam optimization method (Kingma and Ba, 2014).

For regularization, we only apply dropout (Srivastava et al., 2014) before the output layer. We also considered $L_2$ regularization but did not find a consistent improvement in preliminary experiments.

### 3.4 Word embeddings

The first layer of the CNN involves mapping words in the input to dense, low-dimensional vectors. These word embeddings are critically important as they represent the "meaning" of the words in the model, e.g. how similar one word is to another. Although it is possible to learn these embeddings from scratch (i.e. random initialization) during the normal training process, recent studies have shown that it is effective to use embeddings that have been separately induced on large, unannotated corpora (Collobert et al., 2011; Kim, 2014). Work in biomedical NLP has further established that word embeddings are domain-dependent: to get the maximal benefit from using pre-trained embeddings for biomedical NLP tasks, the embeddings must be induced using biomedical texts (Stenetorp et al., 2012).

We consider a variety of word embeddings induced using models implemented in the popular `word2vec` package (Mikolov et al., 2013a). First, we use the general-domain Google News vectors

| | Source texts | | Vectors | | | |
| Name | domain | size | words | dim | OOV | Reference |
| --- | --- | --- | --- | --- | --- | --- |
| Google News | General | 100B | 3M | 300 | 31.0% | (Mikolov et al., 2013b) |
| Pyysalo PM | Bio | 3B | 2.3M | 200 | 0.52% | (Pyysalo et al., 2013a) |
| Pyysalo PMC | Bio | 2.5B | 2.5M | 200 | 0.51% | (Pyysalo et al., 2013a) |
| Pyysalo PM+PMC | Bio | 5.5B | 4M | 200 | 0.49% | (Pyysalo et al., 2013a) |
| Pyysalo Wiki+PM+PMC | General and bio | 7.5B | 5.4M | 200 | 0.53% | (Pyysalo et al., 2013a) |
| Chiu win-2 | Bio | 2.7B | 2.2M | 200 | 0.49% | (Chiu et al., 2016) |
| Chiu win-30 | Bio | 2.7B | 2.2M | 200 | 0.49% | (Chiu et al., 2016) |

Table 3: Word vectors

also applied by Kim (2014).[2]. Second, we evaluate three sets of word vectors induced on various combinations of PubMed (PM), PMC and Wikipedia texts by Pyysalo et al. (2013a).[3] Finally, we consider two variants of PubMed-based vectors introduced by Chiu et al. (2016).[4] The properties of these word vectors are detailed in Table 3. Note that unlike the other properties, the out-of-vocabulary rate (OOV) is not a characteristic of the word vectors alone, but the ratio of words in the task training data that do not appear in the work vectors. The high OOV rate for the Google News vectors is due primarily to removal of stopwords, punctuation, and numbers (see also Section 4.1).

## 3.5 Experimental Setup

Classifier performance is evaluated using the standard precision, recall, and F-score metrics as well as with the area under the receiver operating characteristic curve (AUC). Unlike precision and F-score, AUC is invariant to the positive/negative class distribution. AUC is also more sensitive in summarizing performance over all possible classification thresholds and eliminates the need to pick a specific threshold for evaluation. AUC is therefore recommended for evaluating imbalanced datasets (Zhang and Wallace, 2015). As the dataset is comparatively small and the number of positive examples in particular is very limited for many labels, the random factors in CNN initialization and training can have a substantial effect on the resulting model. To address this issue, we systematically repeated each CNN experiment 10 times and report the mean of the evaluation results.[5] To address overfitting in the CNN, we apply a form of early stopping, testing only the model that achieved the highest results on the development set. In the development experiments, we correspondingly report the highest f-score and AUC from any epoch.

## 4 Results

In the following, we first summarize results from adapting the basic CNN to the task using the development data, and then present the comparative results on the test set.

### 4.1 Development results

We considered a range of modifications to the basic CNN model to better adapt it to biomedical domain text classification in general and the specific task studied in this work in particular. Of these modifications, evaluation on the development set identified three that appeared to have beneficial effects on performance: oversampling to address the class imbalance, using in-domain word vectors, and adjusting the filter sizes to the task. We next briefly describe these modifications and the associated results.

**Oversampling** The dataset is highly biased, with negative examples outnumbering positives more than 10-fold for a number of the labels (Table 1). Standard training on such data is likely to result in models with high precision, low recall, and thus comparatively low F-scores. Addressing this, we oversample the positive examples in the training set with replacement so that their number matches that of the negatives. This modification increased the average F-score on the development set from 85.3% to 86.1%. As expected, the effect on the distribution-independent AUC metric was more limited, improving from 97.3% to 97.5% with oversampling.

---

[2] Available from https://code.google.com/archive/p/word2vec/
[3] Available from http://bio.nlplab.org/
[4] Available from https://github.com/cambridgeltl/BioNLP-2016
[5] As SVM optimization is convex, repetitions are unnecessary.

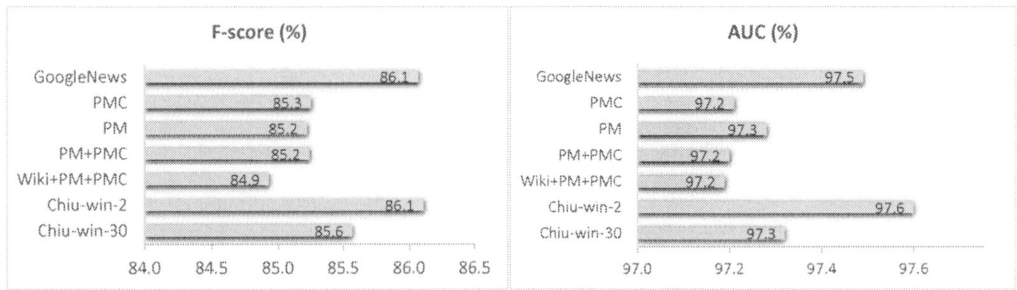

Figure 2: Embedding performance (macro-averaged) on the development dataset

**Word embeddings**   As discussed in Section 3.4, the word vectors used to initialize the embedding layer of the network can have a significant effect on performance. We trained the models using each of the word vectors shown in Table 3 with oversampling (see above) and evaluated development set performance using the maximum F-score and AUC metrics. The results are summarized in Figure 2. Surprisingly, we find that the general domain Google News vectors give very competitive performance despite their high out-of-vocabulary rate (see Table 3), outperforming all in-domain vectors with the exception of the window size 2 word vectors of Chiu et al. (2016). Even these biomedical word vectors only show very modest advantage over the Google News vectors for AUC. In the last development set experiments below and the final test set experiments, we apply the PubMed-based vectors induced with window size 2 from Chiu et al. that were shown to give the best results here.

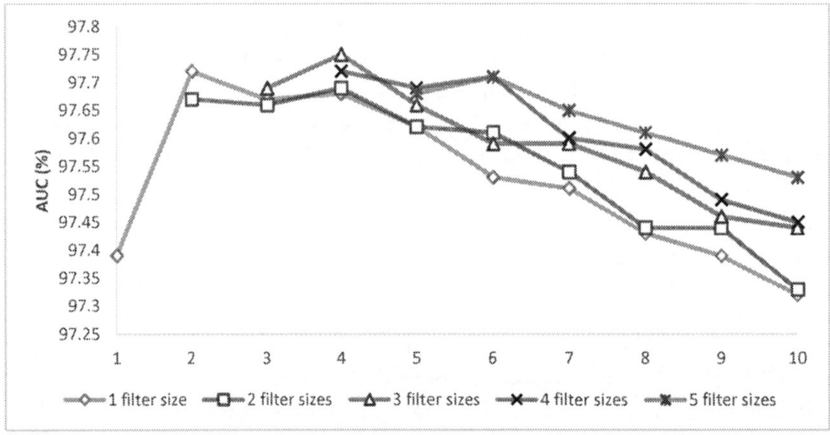

Figure 3: Macro-average AUC with respect to a varying number of filter sizes. Each point on the graph represents the maximum size of filter used (e.g. for *2 filter sizes*, performance with filters of sizes 2 and 3 is plotted at 3).

**Filter sizes**   We experiment with varying the number of filter sizes in the convolutions. The base model of Kim (2014) uses three filter sizes (3,4,5); as part of our hyperparameter search, we investigated what happens to the performance (AUC) with respect to varying filter sizes (1–10) and numbers of filter sizes (1–5), while keeping the total number of filters constant at 300 and filter sizes are ordered consecutively. Figure 3 shows that performance generally falls when increasing the filter size, and the best performance is achieved using three filters of sizes (2,3,4). Another important observation is that the variation in performance is not very substantial, implying that the model is fairly robust to the specific setting of this parameter.

## 4.2 Test results

|  | SVM | | CNN | |
| Hallmark | BoW | Rich | Base | Tuned |
| --- | --- | --- | --- | --- |
| Sustaining proliferative signaling | **70.0%** | 67.4% | 66.3% | 67.9% |
| Evading growth suppressors | 53.5% | 65.3% | 66.7% | **71.5%** |
| Resisting cell death | 75.9% | 82.7% | **86.9%** | 86.7% |
| Enabling replicative immortality | 73.1% | 90.9% | 91.2% | **91.5%** |
| Inducing angiogenesis | 73.9% | **85.7%** | 74.8% | 79.4% |
| Activating invasion and metastasis | 72.5% | 72.7% | 82.0% | **82.6%** |
| Genomic instability and mutation | 71.2% | 69.2% | 72.2% | **81.7%** |
| Tumor promoting inflammation | 69.9% | 76.6% | 81.6% | **84.2%** |
| Cellular energetics | 78.1% | 85.7% | 76.6% | **88.3%** |
| Avoiding immune destruction | 54.3% | 71.8% | 67.7% | **75.8%** |
| **Average** | 69.2% | 76.8% | 76.6% | **81.0%** |

Table 4: Comparison of test results using F-score

|  | SVM | | CNN | |
| Hallmark | BoW | Rich | Base | Tuned |
| --- | --- | --- | --- | --- |
| Sustaining proliferative signaling | 88.6 | 88.9 | **92.1%** | 91.0% |
| Evading growth suppressors | 87.9 | 91.7 | 94.8% | **96.4%** |
| Resisting cell death | 92.4 | 95.5 | 97.1% | **97.7%** |
| Enabling replicative immortality | 92.4 | 97.4 | **99.8%** | 99.5% |
| Inducing angiogenesis | 94.7 | 98.4 | 97.9% | **99.1%** |
| Activating invasion and metastasis | 96.0 | 94.0 | 97.8% | **98.2%** |
| Genomic instability and mutation | 92.5 | 91.7 | 95.8% | **97.0%** |
| Tumor promoting inflammation | 92.7 | 95.9 | **98.3%** | 98.1% |
| Cellular energetics | 99.1 | **99.6** | 99.5% | **99.6%** |
| Avoiding immune destruction | 94.6 | 96.1 | 97.8% | **99.1%** |
| **Average** | 93.1 | 94.9 | 97.1% | **97.6%** |

Table 5: Comparison of test results using AUC

The results of the evaluation on the test data are shown in Table 4 for F-score and 5 for AUC. Overall, both metrics agree that the SVM with bag-of-words features has the lowest performance, and the CNN tuned to the task the highest. As could be expected, the SVM with rich features outperforms the base CNN in terms of F-score; however, the latter, generic model achieves a notably higher AUC than the SVM, suggesting that the slight advantage of the former for F-score may be due in part to a better position of the decision boundary.

The CNN tuned to the task achieves the highest performance on average by both metrics, and further has the highest performance for 7/10 individual classification tasks in terms of both F-score and AUC, outperforming the previous state-of-the-art on this dataset.

## 5 Discussion

Our evaluation contrasts methods separated by two methodological divides: discrete, interpretable, hand-engineered features vs. continuous, opaque, automatically learned features for one, and convex optimization vs. gradient descent in a complex landscape with many local minima for the other. The choice between the SVM representing the former choices and the CNN representing the latter is not necessarily only a question of which performs better, but also of methodological fit, both to the broader machine learning framework and for the practitioners applying the approach.

A key point of interest in neural methods is feature learning, i.e. their capacity to learn complex models with minimal manual effort in feature engineering. As shown again in our experiments, a CNN taking only document text and word embeddings induced from unlabeled text as input can outperform an SVM with extensive manually engineered features derived from sources such as syntactic analysis and named entity recognition. While the 3-4% point differences in AUC and F-score are positive results in favor of the CNN, the relative simplicity and generality of the model is arguably a greater advantage supporting the choice of the CNN over the feature-rich SVM — indeed, one might well argue that the most interesting of our results is that the basic general CNN without any task or domain adaptation only narrowly loses to the SVM in F-score, and outperforms it in terms of AUC. The CNNs can be more readily adapted to other tasks and carry much fewer technical requirements: while the SVM system of Baker et al. (2016) requires running separate tools for lemmatization, parsing, and named entity tagging in addition to the machine learning method, the CNN has no such external dependencies.

For practitioners familiar with SVMs and domain NLP tools, it should be noted that the potential shift to neural methods is not without its own issues. As detailed by Zhang and Wallace (2015), even the simple CNN model considered here comes with a potentially overwhelming number of hyperparameters and related modeling and optimization choices, many of which have task-specific optima, and the cost of training and evaluating large numbers of model variants can be prohibitive even on modern GPU-based systems.[6] For machine learning researchers used to working with convex optimization problems, the random elements involved in training neural network models can also be a source of frustration, and the need to account for variance from network initialization and training also imposes additional computational costs.

Nevertheless, we believe that the simplicity, performance and rich potential for extension and further development of CNNs are more than sufficient to motivate further research on this class of models also in biomedical NLP and anticipate that many domain text classification tasks will see new state of the art results through the use of this class of neural networks.

## 6   Conclusions

In this study, we have considered the application of convolutional neural networks to the biomedical domain text classification task of identifying the hallmarks of cancer associated with publication abstracts.

Using a recently introduced corpus, we demonstrated that a CNN model taking only the document text and word representations induced from unannotated general-domain text as input can achieve competitive performance with a previously proposed SVM-based state-of-the-art classifier with rich manually engineered features including syntactic analyses and named entity recognition outputs. We further adapted the CNN to the task by oversampling positive examples to counteract the class bias, using word vectors induced on biomedical domain text, and optimizing the filter sizes through evaluation on the development set. The adapted model was shown to outperform the SVM, establishing a new state-of-the-art result for this dataset.

We make all of the resources involved in this study available under open source and open data licenses from `https://cambridgeltl.github.io/cancer-hallmark-cnn/`.

## Acknowledgements

The first author is funded by the Commonwealth Scholarship and the Cambridge Trust. This work is supported by Medical Research Council grant MR/M013049/1 and the Google Faculty Award.

## References

Simon Baker, Ilona Silins, Yufan Guo, Imran Ali, Johan Högberg, Ulla Stenius, and Anna Korhonen. 2016. Automatic semantic classification of scientific literature according to the hallmarks of cancer. *Bioinformatics*, 32(3):432–440.

Billy Chiu, Gamal Crichton, Anna Korhonen, and Sampo Pyysalo. 2016. How to train good word embeddings for biomedical NLP. In *Proceedings of BioNLP*.

François Chollet. 2015. Keras. `https://github.com/fchollet/keras`.

Ronan Collobert, Jason Weston, Léon Bottou, Michael Karlen, Koray Kavukcuoglu, and Pavel Kuksa. 2011. Natural language processing (almost) from scratch. *Journal of Machine Learning Research*, 12(Aug):2493–2537.

---

[6]We performed our CNN experiments on the Cambridge high performance computing cluster using NVIDIA K20 GPUs. Although individual epochs completed in 5 seconds on average and model training times were a few minutes at most, the repetitions and parameter grid involved training over 5000 models, and the total training time for the experiments was approximately 150 GPU-hours, not including preliminary and discarded experiments not reported here.

Mary E. Doland. 2014. Capturing cancer initiating events in OncoCL, a cancer cell ontology. In *AMIA Jt Summits Transl Sci*.

Rachel Ginn, Pranoti Pimpalkhute, Azadeh Nikfarjam, Apurv Patki, Karen OConnor, Abeed Sarker, Karen Smith, and Graciela Gonzalez. 2014. Mining twitter for adverse drug reaction mentions: a corpus and classification benchmark. In *Proceedings of BioTxtM 2014*.

Douglas Hanahan and Robert A Weinberg. 2000. The hallmarks of cancer. *cell*, 100(1):57–70.

Douglas Hanahan and Robert A Weinberg. 2011. Hallmarks of cancer: the next generation. *cell*, 144(5):646–674.

Yoon Kim. 2014. Convolutional neural networks for sentence classification. *arXiv preprint arXiv:1408.5882*.

Diederik Kingma and Jimmy Ba. 2014. Adam: A method for stochastic optimization. *arXiv preprint arXiv:1412.6980*.

Yann LeCun and Yoshua Bengio. 1995. Convolutional networks for images, speech, and time series. *The handbook of brain theory and neural networks*, 3361(10):1995.

Hee-Jin Lee. 2014. Oncosearch: cancer gene search engine with literature evidence. *Nucl. Acids Res.*

Nut Limsopatham and Nigel Collier. 2016. Modelling the combination of generic and target domain embeddings in a convolutional neural network for sentence classification. In *Proceedings of BioNLP'16*, page 136.

Haibin Liu, Tom Christiansen, William A Baumgartner Jr, and Karin Verspoor. 2012. BioLemmatizer: a lemmatization tool for morphological processing of biomedical text. *J. Biomedical Semantics*, 3:3.

Tomas Mikolov, Kai Chen, Greg Corrado, and Jeffrey Dean. 2013a. Efficient estimation of word representations in vector space. In *Proceedings of ICLR*.

Tomas Mikolov, Ilya Sutskever, Kai Chen, Greg S Corrado, and Jeff Dean. 2013b. Distributed representations of words and phrases and their compositionality. In *Proceedings of NIPS*, pages 3111–3119.

Fabian Pedregosa, Gaël Varoquaux, Alexandre Gramfort, Vincent Michel, Bertrand Thirion, Olivier Grisel, Mathieu Blondel, Peter Prettenhofer, Ron Weiss, Vincent Dubourg, et al. 2011. Scikit-learn: Machine learning in python. *Journal of Machine Learning Research*, 12(Oct):2825–2830.

Sampo Pyysalo, Filip Ginter, Hans Moen, Tapio Salakoski, and Sophia Ananiadou. 2013a. Distributional semantics resources for biomedical text processing. *Proceedings of LBM*.

Sampo Pyysalo, Tomoko Ohta, and Sophia Ananiadou. 2013b. Overview of the cancer genetics (CG) task of BioNLP Shared Task 2013. In *BioNLP Shared Task 2013 Workshop*.

Laura Rimell and Stephen Clark. 2009. Porting a lexicalized-grammar parser to the biomedical domain. *Journal of biomedical informatics*, 42(5):852–865.

Burr Settles. 2005. ABNER: an open source tool for automatically tagging genes, proteins and other entity names in text. *Bioinformatics*, 21(14):3191–3192.

Aliaksei Severyn and Alessandro Moschitti. 2015. UNITN: Training deep convolutional neural network for twitter sentiment classification. In *Proceedings of SemEval 2015*, pages 464–469.

Nitish Srivastava, Geoffrey E Hinton, Alex Krizhevsky, Ilya Sutskever, and Ruslan Salakhutdinov. 2014. Dropout: a simple way to prevent neural networks from overfitting. *Journal of Machine Learning Research*, 15(1):1929–1958.

Pontus Stenetorp, Hubert Soyer, Sampo Pyysalo, Sophia Ananiadou, and Takashi Chikayama. 2012. Size (and domain) matters: Evaluating semantic word space representations for biomedical text. In *Proceedings of SMBM'12*.

Lin Sun and Anna Korhonen. 2009. Improving verb clustering with automatically acquired selectional preferences. In *Proceedings of the 2009 Conference on Empirical Methods in Natural Language Processing: Volume 2-Volume 2*, pages 638–647. Association for Computational Linguistics.

Ye Zhang and Byron Wallace. 2015. A sensitivity analysis of (and practitioners' guide to) convolutional neural networks for sentence classification. *arXiv preprint arXiv:1510.03820*.

Xiang Zhang, Junbo Zhao, and Yann LeCun. 2015. Character-level convolutional networks for text classification. In *Advances in Neural Information Processing Systems*, pages 649–657.

# Learning Orthographic Features in Bi-directional LSTM for Biomedical Named Entity Recognition


**Nut Limsopatham**  and  **Nigel Collier**
Language Technology Lab
Department of Theoretical and Applied Linguistics
University of Cambridge
Cambridge, UK
{nl347,nhc30}@cam.ac.uk



## Abstract

End-to-end neural network models for named entity recognition (NER) have shown to achieve effective performances on general domain datasets (e.g. newswire), without requiring additional hand-crafted features. However, in biomedical domain, recent studies have shown that hand-engineered features (e.g. orthographic features) should be used to attain effective performance, due to the complexity of biomedical terminology (e.g. the use of acronyms and complex gene names). In this work, we propose a novel approach that allows a neural network model based on a long short-term memory (LSTM) to automatically learn orthographic features and incorporate them into a model for biomedical NER. Importantly, our bi-directional LSTM model learns and leverages orthographic features on an end-to-end basis. We evaluate our approach by comparing against existing neural network models for NER using three well-established biomedical datasets. Our experimental results show that the proposed approach consistently outperforms these strong baselines across all of the three datasets.


## 1   Introduction

Named entity recognition (NER) is one of the first and important stages in a natural language processing (NLP) pipeline. In particular, an NER task is to identify mentions of entities (e.g. persons, locations and organisations) within unstructured text. In biomedical domain, NER tasks are particularly difficult, since the entities of interests are mainly genes, proteins, and chemical substances, which by nature (1) consist of millions of entities, (2) are created continuously, and (3) are non-standardised and can be referred to using different names (e.g. the use of acronyms and polysemy) (Kim et al., 2009; Kim et al., 2004; Smith et al., 2008a).

Traditionally, most of the effective NER approaches are based on machine learning techniques, such as conditional random field (CRF), support vector machine (SVM) and perceptrons (Lafferty et al., 2001; McCallum and Li, 2003; Settles, 2004; Luo et al., 2015; Ju et al., 2011; Ratinov and Roth, 2009; Segura-Bedmar et al., 2015). For instance, Ratinov and Roth (2009) effectively learned a perceptron model using features, including word classes induced using Brown clustering (Liang, 2005), and gazetteer extracted from Wikipedia. Campos et al. (2013) achieved effective performances for several biomedical NER tasks by learning a CRF model using multiple sets of features, including orthographic, morphological, linguistic-based, conjunctions and dictionary-based. However, these approaches rely heavily on feature engineering and domain knowledge (e.g. gazetteers), which are costly to develop. Consequently, they are difficult to be adapted to a new domain, since hand-engineered features are mostly specific to a target domain.

Recent advances in word vector representation (i.e. word embeddings) (Mikolov et al., 2013; Pennington et al., 2014), which represents a word in the form of a low-dimensional vector of real values, allow machine learning approaches to exploit semantic and syntactic information from word vectors, induced






*Proceedings of the Fifth Workshop on Building and Evaluating Resources for Biomedical Text Mining (BioTxtM 2016),*
pages 10–19, Osaka, Japan, December 12th 2016.

from a large dataset, for several NLP tasks, such as NER, part-of-speech (POS) tagging, sentiment analysis and concept normalisation (Collobert et al., 2011; Turian et al., 2010; Limsopatham and Collier, 2016a; Limsopatham and Collier, 2016b; Limsopatham and Collier, 2015). For example, Collobert et al. (2011) effectively used word embeddings as inputs of a feed-forward neural network for sequence labelling tasks, such as NER and POS tagging. Turian et al. (2010) learned a CRF model using word embeddings as input features for NER and chunking tasks. In the biomedical domain, Chiu et al. (2016) investigated the the use of different word embeddings in a feed-forward neural network for biomedical NER tasks. However, when using with word embedding features, traditional features (e.g. orthography and gazetteers) have shown to further improve the performance of an NER system (Segura-Bedmar et al., 2015; Turian et al., 2010; Huang et al., 2015).

In this work, we investigate a novel approach that allows an end-to-end neural network system for biomedical NER to explicitly learn and leverage orthographic features. Our approach is based on bi-directional long short-term memory (LSTM) (Hochreiter and Schmidhuber, 1997) that learns to identify named entities in a sentence using both word and character embeddings as inputs. In particular, for each input sentence, we propose to generate and feed *an orthographic sentence* into a bi-directional LSTM to enable the model to explicitly learn orthographic features. We evaluate our proposed approach using three different well-established biomedical test collections, including the BioCreative II Gene Mention task corpus (BC2) (Smith et al., 2008b), the BioNLP 2009 shared task on event extraction (BioNLP09) (Kim et al., 2009) and the NCBI disease corpus (NCBI) (Doğan et al., 2014). Our experimental results show that the proposed approach consistently outperforms existing effective baselines in term of the f1-score measure.

The main contributions of this paper are three-folds:

1. We investigate the use of both word and character embeddings in bi-directional LSTM for biomedical NER tasks.

2. We propose a novel approach that enables bi-directional LSTM to automatically learn and leverage orthographic features without requiring feature engineering.

3. We thoroughly evaluate our proposed approach using three different standardised datasets for biomedical NER.

The remainder of this paper is organised as follows. In Section 2, we discuss related work and position our paper in the literature. In Section 3, we introduce our approach to learn and leverage orthographic features in bi-directional LSTM for biomedical NER. In Sections 4 and 5, we describe our experimental setup and empirically evaluate our approach, respectively. Section 6 provides concluding remarks.

## 2  Related Work

Biomedical NER, which aims to identify chunks of text mentioning specific entities of interest, is one of the fundamental biomedical text mining tasks. Due to the rapid growth of the number of biomedical documents, an automatic text mining system is needed to extract knowledge from the vast amount of data. Different from a general domain (e.g. newswire) where entities of interest are mainly places, persons and organisations (Tjong Kim Sang and De Meulder, 2003), entities that biomedical NER tasks focus on are, for example, genes, proteins, DNA and RNA. Existing studies (e.g. (Zhou et al., 2004; Fukuda et al., 1998; Liu et al., 2002)) showed that unique characteristics of biomedical text made NER a challenging task, such that existing NER approaches used in a general domain might not be effective. For example, Zhou et al. (2004) found that the names of many of biomedical entities were typically long (i.e. containing at least four words). In addition, the use of non-standardised naming conventions and abbreviation poses a significant challenge in biomedical NER (Smith et al., 2008a). For instance, 'cholesterol' can also be referred as '(3)-cholest-5-en-3-ol', '(3beta)-cholest-5-en-3-ol', '(3b)-cholest-5-en-3-ol', '5-Cholesten-3beta-ol' or '5-Cholesten-3b-ol'.

Machine learning-based approaches for NER have shown to achieve state-of-the-art performances for both general and biomedical domains. Conditional random field (CRF) is one of the most effective

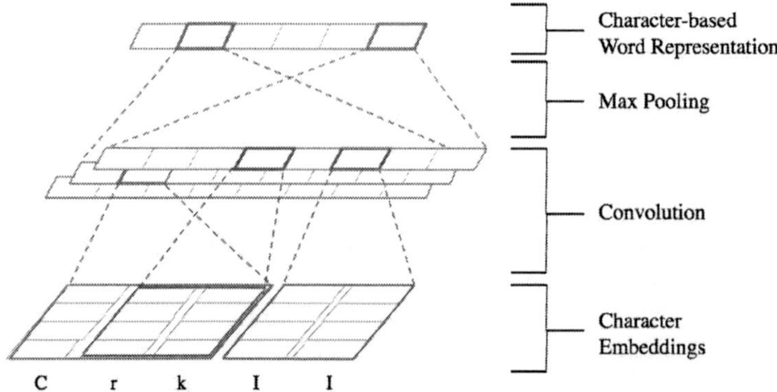


Figure 1: Our CNN architecture for learning word representation from character embeddings.

approaches used in NER tasks (Lafferty et al., 2001; McCallum and Li, 2003; Settles, 2004). Specifically, CRF is based on an undirected statistical graphical model that aims to learn a latent structure of an input sequence. Examples of effective biomedical NER tools that are based on CRF are ABNER (Settles, 2005), BANNER (Leaman et al., 2008) and Gimli (Campos et al., 2013). However, the performance of these CRF-based tools heavily depend on hand-crafted features, such as orthographic and contextual features (Bikel et al., 1999; Collier et al., 2000), which are task-specific and costly to develop. For example, Segura-Bedmar et al. (2015) manually created orthographic features, such as upperInitial (i.e. whether a given word begins with an upper-case character and then follows by any lower-case characters) and allCaps (i.e. whether all characters in a given word are upper-case), when learning a CRF model for drug name recognition. In this work, we investigate an automatic approach that could automatically induce orthographic features for biomedical named entity recognition.

Recently, neural network-based approaches have been effectively used for NER tasks. For example, Collobert et al. (2011) used a feed-forward neural network to effectively identify entities in a newswire corpus (Tjong Kim Sang and De Meulder, 2003) by classifying each word using contexts within a fixed number of surrounding words. Ma and Hovy (2016) and Lample et al. (2016) effectively used both character and word embeddings in a bi-directional LSTM for NER tasks, such as CoNLL03 (Tjong Kim Sang and De Meulder, 2003). Huang et al. (2015) combined hand-crafted features with bi-directional LSTM to further improve the performance. Chiu and Nichols (2016) achieved state-of-the-art performances by modelling both character and word embeddings before combining with hand-crafted features. Nevertheless, the studies of neural network models for biomedical NER tasks are limited. For instance, Chiu et al. (2016) investigated the use of the model of Collobert et al. (2011) with different word embeddings for the BioCreative II Gene Mention task (Smith et al., 2008b) and the JNLPBA task (Kim et al., 2004). In this work, we propose a novel end-to-end neural network model that can learn and leverage orthographic features, which are traditional domain-knowledge features widely used for NER tasks, without requiring any feature engineering.

## 3   Learning Orthographic Features in Bi-directional LSTM

In this section, we introduce our neural network architecture based on bi-directional LSTM for learning and leveraging orthographic features. In particular, our bi-directional LSTM model is composed of (1) character-based word representation, which induces a representation of a word from a character level using a convolutional neural network (CNN) (Section 3.1), (2) word representation, where any pre-trained word embeddings can be used (Section 3.2) and (3) bi-directional LSTM that learns to induce and leverage orthographic features when identifying named entities (Section 3.3).

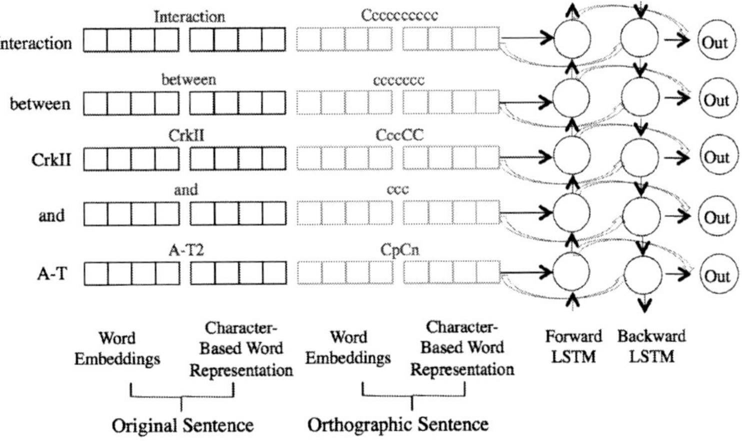


Figure 2: Our bi-directional LSTM for named entity recognition.

## 3.1 Character-based Word Representation

To learn a word representation from a character level, we use CNN to extract important features from character embeddings of a given word, as shown in Figure 1. In particular, we firstly represent a given word of length $l$ characters (padded where necessary) using a word matrix $\mathbf{M} \in \mathbb{R}^{d \times l}$:

$$
\mathbf{M} = \begin{bmatrix} | & | & | & & | \\ \mathbf{x}_1 & \mathbf{x}_2 & \mathbf{x}_3 & ... & \mathbf{x}_l \\ | & | & | & & | \end{bmatrix}
\tag{1}
$$

where each column of $\mathbf{M}$ is the $d$-dimensional vector (i.e. character embedding) $\mathbf{x}_i \in \mathbb{R}^d$ of each character in the given word, which are initialised randomly.

Next, we apply a convolution operation using a filter $\mathbf{w} \in \mathbb{R}^{d \times h}$ to a window of $h$ characters. The filter $\mathbf{w}$ is convolved over the sequence of characters in the word matrix $\mathbf{M}$ to create a feature matrix $\mathbf{C}$. Indeed, each feature $c_i$ in $\mathbf{C}$ is extracted from a window of words $\mathbf{x}_{i:i+h-1}$, as follow:

$$
c_i = f(\mathbf{w} \cdot \mathbf{x}_{i:i+h-1} + b)
\tag{2}
$$

where $f$ is an activation function (such as tanh) and $b \in \mathbb{R}$ is a bias. Note that multiple filters can be used to extract multiple features. In this work, we use 200 filters, each of which has window size $h = 3$.

This convolution operation enables the learning of patterns of characters in words. In order to capture the most important features, max pooling (Collobert et al., 2011) is applied to take the maximum value of each row in the matrix $\mathbf{C}$:

$$
\mathbf{c}_{max} = \begin{bmatrix} max(\mathbf{C}_{1,:}) \\ \vdots \\ max(\mathbf{C}_{d,:}) \end{bmatrix}
\tag{3}
$$

The $\mathbf{c}_{max}$ vector will later be used as a character-based word representation in bi-directional LSTM, since it captures important features of a given word.

## 3.2 Word Representation

We also use pre-trained word embeddings as inputs of bi-directional LSTM, since existing work (e.g. (Mikolov et al., 2013; Pyysalo et al., 2013; Pennington et al., 2014)) has shown that these embeddings could capture semantic and syntactic information of words.

| Input Sentence | Orthographic Sentence |
|---|---|
| interaction between CrkII and A-T2 | cccccccccccc ccccccc CccCC ccc CpCn |
| Prognosis of asymptomatic multiple myeloma. | Ccccccccc cc ccccccccccccc ccccccccc ccccccccp |
| activation of 3-hydroxy-3-methylglutaryl | cccccccccc cc npcccccccccpnpcccccccccccccccc |
| Modification of dopamine D2 receptor activity | Ccccccccccccc cc ccccccccc Cn ccccccccc ccccccccc |
| G alpha i2 and G alpha i2 | C ccccc cn ccc C ccccc cn |
| TPA induction of FGF-BP gene | CCC ccccccccc cc CCCpCC cccc |
| KAP-1 mediated repression in vivo | CCCpn ccccccccc cccccccccc cc cccc |

Table 1: Examples of biomedical sentences and their corresponding orthographic sentence.

| | BC2 | BioNLP09 | NCBI |
|---|---|---|---|
| Target entities | Genes | Bio-molecular events | Diseases |
| Type of data | MEDLINE abstracts | MEDLINE abstracts | PubMed articles |
| Number of documents for training | 201 | 1,436 | 8,662 |
| Number of documents for development | 488 | 995 | 2,872 |
| Number of documents for testing | 58 | 2,200 | 1,036 |

Table 2: The three datasets used to evaluate our proposed approach.

### 3.3 Bi-directional LSTM

We use bi-directional LSTM to learn to identify named entities in a sentence, because it can capture past (from the previous words) and future (from the next words) information effectively (Huang et al., 2015; Dyer et al., 2015). In addition, LSTM has shown to capture long-distance dependencies more effectively than a vanilla recurrent neural networks (RNNs), since it can cope with the gradient vanishing/exploding problems better (Dyer et al., 2015; Bengio et al., 1994).

To enable bi-directional LSTM to learn orthographic features, we create an orthographic pattern of the input sentence (denoted, *the orthographic sentence*). Specifically, given an input sentence (e.g. 'interaction between CrkII and A-T2'), we generate *an orthographic sentence* (e.g. 'cccccccccccc ccccccc CccCC ccc CpCn') by using a set of simple rules, where each of the upper-case characters, lower-case characters, numbers and punctuations, are replaced with $C$, $c$, $n$ and $p$, respectively. Examples of orthographic sentences are shown in Table 1. The orthographic sentence enables bi-directional LSTM to learn orthographic features automatically.

Next, as shown in Figure 2, given an input sentence and its orthographic sentence, we firstly extract both word embeddings (i.e. word representation) and character-based word representation corresponding to each word in the input sentence and the orthographic sentence, by using the approaches described in Sections 3.1 and 3.2[1]. Then, we concatenate word representations associated to the same words and sequentially feed them into bi-directional LSTM to model the contextual information of each word. Finally, at the output layer, we follow Huang et al. (2015) and optimise the CRF log-likelihood, which aims to maximise the likelihood of labelling the whole sentence correctly, by modelling the interactions between two successive labels using the Viterbi algorithm.

## 4 Experimental Setup

### 4.1 Datasets

To evaluate our proposed approach, we use three different well-estabished biomedical NER datasets, which are the BioCreative II Gene Mention task corpus (BC2) (Smith et al., 2008b), the BioNLP 2009 shared task on event extraction (BioNLP09) (Kim et al., 2009) and the NCBI disease corpus (NCBI) (Doğan et al., 2014), respectively. Table 2 shows the information of the three datasets. Firstly, the BC2 dataset consists of 20,000 sentences extracted from MEDLINE abstracts (15,000 sentences for

---

[1] Note that we use separated set of word and character embeddings for the input sentence and the orthographic sentence.

training and 5,000 sentences for testing), where the task is to annotate the mentions of genes. In order to create a development set, we randomly split the original 15,000 training sentences into 10,000 and 5,000 training and development sentences. Secondly, the BioNLP09 dataset is composed of 7,449, 1,450 and 2,447 sentences for training, development and testing, respectively. The target entities are bio-molecular events. Thirdly, the NCBI dataset contains more than 6,000 sentences from 793 PubMed articles (593, 100 and 100 articles for training, development and testing, respectively). The task aims to identify mentions of diseases in a given sentence.

## 4.2  Evaluation Measures

We evaluate the performance on the three biomedical NER tasks in terms of f1-score, precision and recall measures:

$$f1\text{-}score = 2 \cdot \frac{precision \cdot recall}{precision + recall}, \tag{4}$$

$$precision = \frac{TP}{TP + FP}, \tag{5}$$

$$recall = \frac{TP}{TP + FN}, \tag{6}$$

where $TP$ (true positive) is the number of named entity chunks that are correctly identified, $FP$ (false positive) is the number of chunks that are mistakenly identified as entities, and $FN$ (false negative) are the number of named entity chunks that are not identified.

## 4.3  Embeddings

### 4.3.1  Word Embeddings

As discussed in Section 3.2, our approach uses word embeddings as inputs when learning an NER model. We use pre-trained word embeddings of Moen et al. (2013), which are publicly available. In particular, the embeddings consists of 200-dimensional vectors of 5.4 million unique words, which are induced from a combined collection of PubMed, PMC and Wikipedia texts using the Skip-gram model from the word2vec tool (Mikolov et al., 2013). For the words that do not exist in the pre-trained embeddings, we use a vector of random values sampled from $[-\sqrt{\frac{3}{dim}}, +\sqrt{\frac{3}{dim}}]$ where $dim$ is the dimension of embeddings as suggested by He et al. (2015).

We use a separated word embeddings for words in the orthographic sentences. In particular, for each word we use a 200-dimensional randomly generated vector, where each dimension is also uniformly sampled from $[-\sqrt{\frac{3}{dim}}, +\sqrt{\frac{3}{dim}}]$.

### 4.3.2  Character Embeddings

For both input sentence (i.e. original sentence) and orthographic sentence, we use 30-dimensional character embeddings for representing each character when inducing the character-based word representation (Equation (1) in Section 3.1). In particular, we initialise the character embeddings with uniform samples from $[-\sqrt{\frac{3}{dim}}, +\sqrt{\frac{3}{dim}}]$. Importantly, we have a separated embedding for each set of characters in the input and orthographic sentences.

## 4.4  Parameter Optimisation

Parameter optimisation is done by mini-batch stochastic gradient descent (SGD) with batch size 50. In particular, the stochastic gradient descent with back-propagation is performed using Adadelta update rule (Zeiler, 2012). Note that we also fine-tune both word and character embeddings by allowing their weights to be modified when performing gradient updates. To reduce the effects of gradient exploding, we follow Pascanu et al. (2013) and use a gradient clipping of 5.0.

To mitigate overfitting, we apply $L_2$ regularisation on the weight vectors, as well as applying dropout (Srivastava et al., 2014) with dropout rate 0.5 for all of the layers in our model. In addition, we use early stopping (Giles, 2001) based on the performance achieved on the development sets.

| Approach | BC2 | | | BioNLP09 | | | NCBI | | |
|---|---|---|---|---|---|---|---|---|---|
| | F1-score | Precision | Recall | F1-score | Precision | Recall | F1-score | Precision | Recall |
| FeedForward | 66.13 | 76.43 | 58.28 | 76.83 | 77.25 | 76.42 | 73.55 | 72.05 | 75.12 |
| BiLSTM | 69.54 | 74.25 | 65.39 | 80.49 | 85.64 | 75.93 | 75.37 | 77.53 | 73.33 |
| CNN-BiLSTM (Char-only) | 79.98 | 81.85 | 78.20 | 85.11 | 87.54 | 82.81 | 82.70 | 83.00 | 82.40 |
| CNN-BiLSTM | 80.25 | 80.75 | **79.76** | 86.54 | 88.90 | 84.31 | 84.19 | 84.33 | **84.06** |
| ORTH-CNN-BiLSTM | **80.58** | **83.01** | 78.28 | **87.06** | **88.91** | **85.29** | **84.26** | **86.67** | 81.98 |

Table 3: Performances in terms of f1-score, precision and recall of our proposed approach and the baselines on the BC2, BioNLP09 and NCBI datasets.

## 4.5 Baselines

We compare our approach with four different baselines, which do not use any hand-engineered features:

1. *FeedForward*: A simple feed-forward neural network model similar to Collobert et al. (2011) with the context window size of 5 and the pre-trained word embeddings described in Section 4.3.1.

2. *BiLSTM*: A bi-directional LSTM model similar to the proposed model in Section 3, excepting that the orthographic sentence and the character-based word representation are discarded from the model. This baseline is similar to the model of Huang et al. (2015) when hand-crafted features are not taken into account.

3. *CNN-BiLSTM (Char-only)*: A bi-directional LSTM model similar to the proposed model in Section 3, excepting that the orthographic sentence and the word embeddings are discarded from the model.

4. *CNN-BiLSTM*: A bi-directional LSTM model similar to the model in Section 3, excepting that the orthographic sentence is not taken into account by the model.

## 5   Experimental Results

In this section, we compare the performance of our approach for learning and leveraging orthographic features in bi-directional LSTM for biomedical NER (denoted, *ORTH-CNN-BiLSTM*) against the four baselines introduced in Section 4.5. Table 3 compares the performances of our proposed approach with the baselines in terms of f1-score, precision and recall on the three datasets (i.e. BC2, BioNLP09 and NCBI).

From Table 3, we firstly observe that *FeedForward* is the weakest baseline, especially in terms of the f1-score. This is intuitive as feed-forward neural network is a simple model in comparison with bi-directional LSTM that could learn long-distance dependencies from sequences of words. Next, we compare the performance of *BiLSTM* and *CNN-BiLSTM (Char-only)*. Both *BiLSTM* and *CNN-BiLSTM (Char-only)* share a similar architecture for identifying named entities. The only difference is that *BiLSTM* uses pre-trained word embeddings for representing words in a sentence; meanwhile, *CNN-BiLSTM (Char-only)* learns word representation from character embeddings using a convolutional neural network. We observe that *CNN-BiLSTM (Char-only)* achieves better performances than *BiLSTM* in terms of all the three reported measures (i.e. f1-score, precision and recall), across the three datasets. This highlights the importance of the character-based word representation that could help to deal with non-standardised and continuously-growing biomedical vocabularies. Furthermore, we found that *CNN-BiLSTM*, which uses both pre-trained word embeddings and character-based word representation in a bi-directional LSTM model, further improves the f1-score and recall performances on all of the three datasets.

On the other hand, our approach, *ORTH-CNN-BiLSTM*, outperforms all of the baselines on the three datasets. In particular, *ORTH-CNN-BiLSTM* performs better than *CNN-BiLSTM*, which is the most effective baseline, in terms of f1-score and precision for all of the BC2, BioNLP09 and NCBI datasets. Importantly, we observe that our approach for automatically learning orthographic features could effectively boost the performance in term of precision. For example, for the BC2 and NCBI datasets,

*ORTH-CNN-BiLSTM* achieved 83.01% and 86.67% precision, while *CNN-BiLSTM* attains 80.75% and 84.33% precision, respectively.

When analysing the performance of *ORTH-CNN-BiLSTM*, we observe that the induced orthographic features could help to effectively identify complex biomedical entities, such as 'CrkII-23', 'ch-IAP1', 'HC-toxin', 'E.coli manX equivalent', 'cathepsin K', 'IL-2', and 'A-T', that do not appear in the training set by learning from the orthographic patterns of words. This shows the importance of orthographic features in biomedical NER tasks. Importantly, our approach shows a potential of enabling bi-directional LSTM to capture these patterns without resorting to hand-engineered features.

## 6   Conclusions

We have discussed recent advances in neural networks that could enable a machine learning-based NER system to performed effectively in a general domain, such as newswire, without requiring any hand-crafted features. However, the complexity and the continuous growth of biomedical vocabularies make biomedical NER a challenging task. Consequently, biomedical NER systems would require domain knowledge, in the forms of hand-crafted features, to achieve an effective performance. In this work, we investigate an approach that allows bi-directional LSTM to automatically learn and leverage orthographic features, which is one of the key features for biomedical NER. We evaluate our approach by comparing against existing effective end-to-end neural network models for NER. Our experimental results evaluated on three different well-established biomedical NER datasets showed that our approach consistently outperformed the baselines. Importantly, we found that our approach could help to identify named entities that did not appear in the training data by learning the orthographic patterns from similar entities. For future work, we aim to enable neural network models to automatically induce other hand-crafted features, such as gazetteers.

### Acknowledgements

The authors wish to thank funding support from the EPSRC (grant number EP/M005089/1).

### References

Yoshua Bengio, Patrice Simard, and Paolo Frasconi. 1994. Learning long-term dependencies with gradient descent is difficult. *IEEE transactions on neural networks*, 5(2):157–166.

Daniel M. Bikel, Richard Schwartz, and Ralph M. Weischedel. 1999. An algorithm that learns what‘s in a name. *Mach. Learn.*, 34(1-3):211–231, February.

David Campos, Sérgio Matos, and José Luís Oliveira. 2013. Gimli: open source and high-performance biomedical name recognition. *BMC bioinformatics*, 14(1):1.

Jason Chiu and Eric Nichols. 2016. Named entity recognition with bidirectional lstm-cnns. *Transactions of the Association for Computational Linguistics*, 4:357–370.

Billy Chiu, Gamal Crichton, Anna Korhonen, and Sampo Pyysalo. 2016. How to train good word embeddings for biomedical nlp. In *Proceedings of BioNLP16*, page 166.

Nigel Collier, Chikashi Nobata, and Jun-ichi Tsujii. 2000. Extracting the names of genes and gene products with a hidden markov model. In *Proceedings of the 18th Conference on Computational Linguistics - Volume 1*, COLING '00, pages 201–207, Stroudsburg, PA, USA. Association for Computational Linguistics.

Ronan Collobert, Jason Weston, Léon Bottou, Michael Karlen, Koray Kavukcuoglu, and Pavel Kuksa. 2011. Natural language processing (almost) from scratch. *J. Mach. Learn. Res.*, 12:2493–2537, November.

Rezarta Islamaj Doğan, Robert Leaman, and Zhiyong Lu. 2014. Ncbi disease corpus: a resource for disease name recognition and concept normalization. *Journal of biomedical informatics*, 47:1–10.

Chris Dyer, Miguel Ballesteros, Wang Ling, Austin Matthews, and Noah A Smith. 2015. Transition-based dependency parsing with stack long short-term memory. *arXiv preprint arXiv:1505.08075*.

Ken-ichiro Fukuda, Tatsuhiko Tsunoda, Ayuchi Tamura, Toshihisa Takagi, et al. 1998. Toward information extraction: identifying protein names from biological papers. In *Pac symp biocomput*, volume 707, pages 707–718. Citeseer.

Rich Caruana Steve Lawrence Lee Giles. 2001. Overfitting in neural nets: Backpropagation, conjugate gradient, and early stopping. In *Advances in Neural Information Processing Systems 13: Proceedings of the 2000 Conference*, volume 13, page 402. MIT Press.

Kaiming He, Xiangyu Zhang, Shaoqing Ren, and Jian Sun. 2015. Delving deep into rectifiers: Surpassing human-level performance on imagenet classification. In *Proceedings of the IEEE International Conference on Computer Vision*, pages 1026–1034.

Sepp Hochreiter and Jürgen Schmidhuber. 1997. Long short-term memory. *Neural computation*, 9(8):1735–1780.

Zhiheng Huang, Wei Xu, and Kai Yu. 2015. Bidirectional lstm-crf models for sequence tagging. *arXiv preprint arXiv:1508.01991*.

Zhenfei Ju, Jian Wang, and Fei Zhu. 2011. Named entity recognition from biomedical text using svm. In *Bioinformatics and Biomedical Engineering,(iCBBE) 2011 5th International Conference on*, pages 1–4. IEEE.

Jin-Dong Kim, Tomoko Ohta, Yoshimasa Tsuruoka, Yuka Tateisi, and Nigel Collier. 2004. Introduction to the bio-entity recognition task at jnlpba. In *Proceedings of the international joint workshop on natural language processing in biomedicine and its applications*, pages 70–75. Association for Computational Linguistics.

Jin-Dong Kim, Tomoko Ohta, Sampo Pyysalo, Yoshinobu Kano, and Jun'ichi Tsujii. 2009. Overview of bionlp'09 shared task on event extraction. In *Proceedings of the Workshop on Current Trends in Biomedical Natural Language Processing: Shared Task*, pages 1–9. Association for Computational Linguistics.

John Lafferty, Andrew McCallum, and Fernando Pereira. 2001. Conditional random fields: Probabilistic models for segmenting and labeling sequence data. In *Proceedings of the eighteenth international conference on machine learning, ICML*, volume 1, pages 282–289.

Guillaume Lample, Miguel Ballesteros, Sandeep Subramanian, Kazuya Kawakami, and Chris Dyer. 2016. Neural architectures for named entity recognition. In *Proceedings of the 2016 Conference of the North American Chapter of the Association for Computational Linguistics: Human Language Technologies*, pages 260–270, San Diego, California, June. Association for Computational Linguistics.

Robert Leaman, Graciela Gonzalez, et al. 2008. Banner: an executable survey of advances in biomedical named entity recognition. In *Pacific symposium on biocomputing*, volume 13, pages 652–663.

Percy Liang. 2005. *Semi-supervised learning for natural language*. Ph.D. thesis, Massachusetts Institute of Technology.

Nut Limsopatham and Nigel Collier. 2015. Adapting phrase-based machine translation to normalise medical terms in social media messages. In *Proceedings of the 2015 Conference on Empirical Methods in Natural Language Processing*, pages 1675–1680, Lisbon, Portugal, September. Association for Computational Linguistics.

Nut Limsopatham and Nigel Collier. 2016a. Modelling the combination of generic and target domain embeddings in a convolutional neural network for sentence classification. In *Proceedings of the 15th Workshop on Biomedical Natural Language Processing*. Association for Computational Linguistics.

Nut Limsopatham and Nigel Collier. 2016b. Normalising medical concepts in social media texts by learning semantic representation. In *Proceedings of the 54th Annual Meeting of the Association for Computational Linguistics (Volume 1: Long Papers)*, pages 1014–1023, Berlin, Germany, August. Association for Computational Linguistics.

Hongfang Liu, Alan R Aronson, and Carol Friedman. 2002. A study of abbreviations in medline abstracts. In *Proceedings of the AMIA Symposium*, page 464. American Medical Informatics Association.

Gang Luo, Xiaojiang Huang, Chin-Yew Lin, and Zaiqing Nie. 2015. Joint entity recognition and disambiguation. In *Proceedings of the 2015 Conference on Empirical Methods in Natural Language Processing*, pages 879–888, Lisbon, Portugal, September. Association for Computational Linguistics.

Xuezhe Ma and Eduard Hovy. 2016. End-to-end sequence labeling via bi-directional lstm-cnns-crf. In *Proceedings of the 54th Annual Meeting of the Association for Computational Linguistics (Volume 1: Long Papers)*, pages 1064–1074, Berlin, Germany, August. Association for Computational Linguistics.

Andrew McCallum and Wei Li. 2003. Early results for named entity recognition with conditional random fields, feature induction and web-enhanced lexicons. In *Proceedings of the seventh conference on Natural language learning at HLT-NAACL 2003-Volume 4*, pages 188–191. Association for Computational Linguistics.

Tomas Mikolov, Ilya Sutskever, Kai Chen, Greg S Corrado, and Jeff Dean. 2013. Distributed representations of words and phrases and their compositionality. In *NIPS*, pages 3111–3119.

Razvan Pascanu, Tomas Mikolov, and Yoshua Bengio. 2013. On the difficulty of training recurrent neural networks. *ICML (3)*, 28:1310–1318.

Jeffrey Pennington, Richard Socher, and Christopher D Manning. 2014. GloVe: Global vectors for word representation. In *EMNLP*, pages 1532–1543.

S. Pyysalo, F. Ginter, H. Moen, T. Salakoski, and S. Ananiadou. 2013. Distributional semantics resources for biomedical text processing. In *Proceedings of LBM 2013*, pages 39–44.

Lev Ratinov and Dan Roth. 2009. Design challenges and misconceptions in named entity recognition. In *Proceedings of the Thirteenth Conference on Computational Natural Language Learning*, pages 147–155. Association for Computational Linguistics.

Isabel Segura-Bedmar, Victor Suárez-Paniagua, and Paloma Martínez. 2015. Exploring word embedding for drug name recognition. In *SIXTH INTERNATIONAL WORKSHOP ON HEALTH TEXT MINING AND INFORMATION ANALYSIS (LOUHI)*, page 64.

Burr Settles. 2004. Biomedical named entity recognition using conditional random fields and rich feature sets. In *Proceedings of the International Joint Workshop on Natural Language Processing in Biomedicine and its Applications*, pages 104–107. Association for Computational Linguistics.

Burr Settles. 2005. Abner: an open source tool for automatically tagging genes, proteins and other entity names in text. *Bioinformatics*, 21(14):3191–3192.

Larry Smith, Lorraine K Tanabe, Rie Johnson nee Ando, Cheng-Ju Kuo, I-Fang Chung, Chun-Nan Hsu, Yu-Shi Lin, Roman Klinger, Christoph M Friedrich, Kuzman Ganchev, et al. 2008a. Overview of biocreative ii gene mention recognition. *Genome biology*, 9(2):1.

Larry Smith, Lorraine K Tanabe, Rie Johnson nee Ando, Cheng-Ju Kuo, I-Fang Chung, Chun-Nan Hsu, Yu-Shi Lin, Roman Klinger, Christoph M Friedrich, Kuzman Ganchev, et al. 2008b. Overview of biocreative ii gene mention recognition. *Genome biology*, 9(2):1.

Nitish Srivastava, Geoffrey E Hinton, Alex Krizhevsky, Ilya Sutskever, and Ruslan Salakhutdinov. 2014. Dropout: a simple way to prevent neural networks from overfitting. *Journal of Machine Learning Research*, 15(1):1929–1958.

Erik F Tjong Kim Sang and Fien De Meulder. 2003. Introduction to the conll-2003 shared task: Language-independent named entity recognition. In *Proceedings of the seventh conference on Natural language learning at HLT-NAACL 2003-Volume 4*, pages 142–147. Association for Computational Linguistics.

Joseph Turian, Lev Ratinov, and Yoshua Bengio. 2010. Word representations: a simple and general method for semi-supervised learning. In *Proceedings of the 48th annual meeting of the association for computational linguistics*, pages 384–394. Association for Computational Linguistics.

Matthew D Zeiler. 2012. Adadelta: an adaptive learning rate method. *arXiv preprint arXiv:1212.5701*.

Guodong Zhou, Jie Zhang, Jian Su, Dan Shen, and Chewlim Tan. 2004. Recognizing names in biomedical texts: a machine learning approach. *Bioinformatics*, 20(7):1178–1190.

# Building Content-driven Entity Networks for Scarce
# Scientific Literature using Content Information


**Reinald Kim Amplayo, Min Song**
Department of Library and Information Science
Yonsei University
Seoul, Korea
{rktamplayo, min.song}@yonsei.ac.kr


## Abstract


This paper proposes several network construction methods for collections of scarce scientific literature data. We define scarcity as lacking in value and in volume. Instead of using the paper's metadata to construct several kinds of scientific networks, we use the full texts of the articles and automatically extract the entities needed to construct the networks. Specifically, we present seven kinds of networks using the proposed construction methods: co-occurrence networks for author, keyword, and biological entities, and citation networks for author, keyword, biological, and topic entities. We show two case studies that applies our proposed methods: CADASIL, a rare yet the most common form of hereditary stroke disorder, and Metformin, the first-line medication to the type 2 diabetes treatment. We apply our proposed method to four different applications for evaluation: finding prolific authors, finding important bio-entities, finding meaningful keywords, and discovering influential topics. The results show that the co-occurrence and citation networks constructed using the proposed method outperforms the traditional-based networks. We also compare our proposed networks to traditional citation networks constructed using enough data and infer that even with the same amount of enough data, our methods perform comparably or better than the traditional methods.


## 1 Introduction

Large amounts of biomedical data can now be procured in the Internet. One of the more trustworthy source of data is from the scientific community where they do research on specific topics and publish them, which is then made available on the Internet. These vast amounts of data have been used successfully in a lot of areas in biomedicine (Margolis et al., 2014; Marx, 2013; Costa, 2014), from biocuration (Howe et al., 2008) to entity extraction (Rindflesch et al., 2000). In this paper, we focus on the application of the social and knowledge network construction to biomedical data.

One major yet unseen problem is the contradicting problem of *scarce data*. In this paper, we define scarcity in two-folds: lack of value and lack of volume. Lacking in value means that it lacks the necessary information to perform the method. In the case of constructing an author citation network, scarce data may not have the author and citing author information in its metadata. Lacking in volume means that it is not big enough to uncover important knowledge. In the case of constructing an author collaboration network, scarce data may not have enough scale to detect meaningful communities.

Both of these problems in scarcity exist in rare diseases since there are still very few research regarding these diseases. In this paper, we focus on a case study on the research area on Cerebral Autosomal Dominant Arteriopathy with Subcortical Infarcts and Leukoencephalopathy, also known as CADASIL. CADASIL (Chabriat et al., 2009) is the most common form of hereditary stroke disorder, yet is listed as one of the many rare diseases[1]. As of the time of writing, searching for research articles regarding CADASIL in Scopus[2] gives approximately only 1100 documents compared to, for example, the approx-



[1]http://globalgenes.org/rarelist
[2]http://www.ncbi.nlm.nih.gov/pmc/




*Proceedings of the Fifth Workshop on Building and Evaluating Resources for Biomedical Text Mining (BioTxtM 2016)*,
pages 20–29, Osaka, Japan, December 12th 2016.

imately 321 thousand lung cancer-related documents. Using current traditional network construction techniques on the CADASIL data may not work properly. Thus, it is necessary to create an alternative method to handle these kinds of data.

This paper proposes alternative methods to constructing social and knowledge networks to handle scarce data. Instead of using the metadata information, which may not be available, we use the full text of the paper to construct networks. More specifically, instead of using the unavailable author and abstract metadata information of the cited papers, we make use of the sentences where the in-text citations are located (which in this paper we call in-text citation context). Aside from it being able to handle scarce data, it also has some other advantages:

- It can discover **larger communities**, which can be subtopics of the subject at hand, or connections to other subjects which are related to the subject at hand.

- In case of constructing entity co-occurrence networks, it defines a much **clearer polarity** on whether the entities are more significant or less significant because the number of citations received by the entity is also reflected.

- In case of constructing entity citation networks, it makes **use of citation information** extensively. Only the part of the cited paper aimed to cite by the citing paper is included. This is an important distinction because even though the communities become larger and may include other subjects, only the related entities are extracted.

We apply our methods to four different tasks: finding influential authors, finding important biological entities, finding meaningful keywords, and discovering trendy topics. We also present a comparative experimental study on metformin, a drug for type 2 diabetes, which was used as a case study in Ding et al. (2013). We note that these tasks are presented to show comparisons between our proposed methods and the traditional methods in *constructing networks*. The novelty of the paper lies on the construction of entity networks through content-driven approaches.

## 2  Related work

In this section, we describe related research works on traditional social and knowledge networks and on methods that utilized in-text citation context.

After Newman (2001) introduced scientific collaboration networks, it has been used to analyze the patterns (Newman, 2004) and structure (Hou et al., 2008) of scientific collaboration and coauthorship inside a research community. Hou et al. (2008) also used author collaboration networks to identify prolific authors using the centrality measures. A more recent study by Song et al. (2014) used author collaboration networks to detect communities within the field. Interestingly, citation graphs where authors are the nodes are not used as much as compared to author collaboration networks. Author citation graphs have been used to define a scientist's weighting factor (Życzkowski, 2010) and to determine the citation strength of productive and highly cited authors (Ding, 2011). Entity-based networks, such as entity co-occurrence and entity citation networks, have also been constructed manually (Callon et al., 1991; Ding et al., 2001), using a dictionary (Pettigrew and McKechnie, 2001; Plake et al., 2006; Yan et al., 2013), and using a machine learning technique (Ding et al., 2013; Hahm and Song, 2015) to describe and measure the impact of the entity community or the entity itself and to detect the hidden knowledge between two entities.

Since there were enough data to do proper network analysis, all of the past works above used only meta information such as the paper's authors and abstracts. Only a few research works used the citation information, both the in-text citation context and the reference section of the paper (Yin et al., 2011; Jeong et al., 2014). Yin et al. (2011) used the in-text citation contexts to model linkage information to improve the retrieval of biomedical documents. Similar as ours, Jeong et al. (2014) takes the citation information and constructs a content-based co-citation author network. They constructed an author co-citation network that considers the two authors' contents' similarity when adding edges between the two authors. In this paper, on the other hand, we propose a method to the construction of co-occurrence and

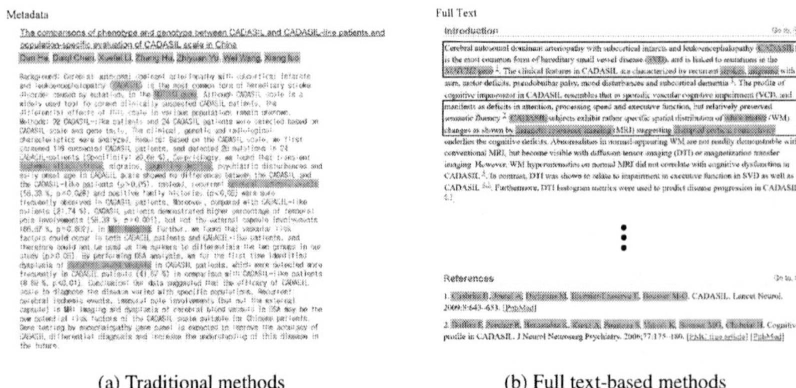

(a) Traditional methods       (b) Full text-based methods

Figure 1: Entity extraction methods

citation networks for scarce data, where if the traditional methods are used to construct the network, network analysis is not possible.

## 3 Network construction

### 3.1 Traditional-based methods

This section introduces our approach to network construction compared to the traditional approaches. Figure 1a shows where the traditional methods extract the nodes or the entities used to construct the network. More generally, traditional methods get their entities from the metadata information. Author-based networks are constructed from the authors (highlighted orange in Figure 1a) of the paper and entity-based networks are constructed from the abstract of the paper. For example, the traditional method in constructing author collaboration networks creates edges between authors extracted from the author lists of the papers. Also, the traditional method in constructing entity-entity citation networks creates edges between entities (highlighted blue in Figure 1a) found in the two abstracts of the papers. The problem lies within the volume and the availability of these metadata information in scarce data. Networks constructed with little data cannot uncover important knowledge.

### 3.2 Full text-based methods

In this paper, we present a network construction method that uses the full texts instead of the available metadata information. Figure 1b shows the location where the full text-based methods extract their entities.

#### 3.2.1 Co-occurrence networks

The full text-based method for constructing a co-occurrence network is similar to the traditional method. The only difference is the location where the entities are extracted. In the case of the author collaboration networks, instead of looking at the authors of the paper, we look at the reference section to extract the authors (highlighted orange in Figure 1b) from the references. One reference citation has one list of authors. A weighted edge is then created between two authors that belongs to one list. Note that if the edge already exists in the graph and another edge between two authors is created, the weight of the existing edge is increased by one. The main advantage of this method is that the constructed collaboration network reflects the number of citations the authors received. This makes it possible to define a much clearer polarity between prolific and non-prolific authors.

In the case of the entity co-occurrence networks, we look at the in-text citations to locate the chunks of text that the citing paper referenced the cited paper. From these chunks of text, we then extract the other entities such as the topic, the keywords, or the biological entities. After extraction, each chunk of text has its list of entities. A weighted edge is then created between two authors that belongs to one list.

### 3.2.2 Citation networks

The main disadvantage of the traditional method in constructing a citation network is the fact that it needs the citation information between two papers as a metadata information. If the author and abstract information of the cited papers are not available, citation networks cannot be constructed using the traditional method.

Our method does not need the citation information as a metadata information. In the case of the author citation networks, we create a directed weighted edge from each of the authors from the metadata information (orange in Figure 1a) to each of the authors from the reference section of the paper (orange in Figure 1b). In the case of the entity citation networks, we create a directed weighted edge from each of the entities extracted from the abstract (blue in Figure 1a) of the paper to each of the entities extracted from all the in-text citation contexts (blue in Figure 1b).

## 3.3 Entity extraction

The full text-based methods need to extract the entities to construct the networks. The authors are gathered from the author metadata information and from the reference section of the full text paper. All the other entities are extracted from the abstract and the in-text citation context.

### 3.3.1 Author extraction

In order to extract the authors from the reference section, it is necessary to take note of the many different styles of citations. Thus, we use an automatic machine learning method to extract the authors from the reference section. We sample a few reference section and manually tag the authors for each reference citation. We then feed them as input for our machine learning model. We use ABNER (Settles, 2005) to create a new linear-chain conditional random field (CRF) based entity extraction where the entity used is only the author. After training, the f1-score of the model is 99.3% with precision of 99.31% and recall of 99.29%. For papers with authors more than 11, we only extract the first 10 and the last author, following the sequence-determines-credit (SDC) and the first-last-author-emphasis (FLAE) approach to author credit contribution (Tscharntke et al., 2007). The author names are then formatted as FN LASTNAME where FN contains the first name initials and LASTNAME is the last name of the author.

### 3.3.2 Bio-entity extraction

There are multiple types of biological entities from diseases and genes to chemicals and proteins. We use PKDE4J (Song et al., 2015), a biological entity extraction text mining system that synthesized the extraction of 127 types of biological entities obtained from the UMLS semantic groups. Out of the two available methods, we make use of the machine learning-based entity extraction. Since the extracted entities are not preprocessed, we do simple preprocessing techniques by removing the non-alphanumeric symbols, removing multiple whitespaces, and lemmatizing the words using Stanford CoreNLP (Manning et al., 2014).

### 3.3.3 Keyword extraction

We also extract keywords from the text automatically by using the rapid automatic keyword extraction (RAKE) algorithm (Rose et al., 2010). RAKE is an unsupervised domain- and language-independent method for extracting keywords by making use of a generated stoplist which makes it usable for different domains and languages. In this paper, we use the SMART English stopword list provided by Salton et al. (1975) as the stoplist. After the extraction, we use the same techniques in Section 3.3.2 to preprocess the extracted keywords.

### 3.3.4 Topic extraction

Topics are extracted using the latent Dirichlet allocation (LDA) topic model (Blei et al., 2003). LDA is a topic modeling technique that infers each document its own topic given the words of each document and two Dirichlet priors $\alpha$ and $\beta$. We set the number of topics to 500 and the number of iterations to 5000. We set the Dirichlet priors $\alpha = 1$ and $\beta = 0.01$. The LDA topic model returns a document-topic distribution. From this distribution, we get the two topics with highest probabilities for each abstract and

Table 1: Dataset and network description

| | | author | bio-entity | keyword | topic |
|---|---|---|---|---|---|
| traditional | nodes | 4,707 | 3,493 | 17,033 | - |
| co-occurrence | edges | 18,948 | 40,386 | 369,818 | - |
| full text-based | nodes | 84,180 | 21,987 | 142,319 | - |
| co-occurrence | edges | 295,066 | 89,298 | 846,269 | - |
| full text-based | nodes | 87,719 | 24,522 | 150,895 | 498 |
| citation | edges | 952,994 | 310,590 | 4,513,469 | 17,603 |

one topic for each citation context. We get two topics for the abstracts because the text is long and might be dealing with multiple topics.

## 4 Experiments

### 4.1 Dataset

We gather our datasets from PubMed Central (PMC). We use the query term *cadasil* to get the papers' author information and abstract from MEDLINE and PubMed Central IDs directly from PMC. Using the PMCIDs, we obtain the full text, excluding the abstract and including the reference section. From the full text, we extract the in-text citation context with the guidance from the reference section. The citation context contains at most 60 tokens: from the in-text citation, thirty tokens to the left or until the end of the paragraph, and thirty tokens to the right or until the end of the paragraph.

Multiple networks are then created using the methods described in Section 3. Table 1 shows the statistics of the networks created. There are a total of 10 networks: three traditional co-occurrence networks, three full text-based co-occurrence networks, and four full text-based citation networks. Since the paper's citation information is not available, citation networks using the traditional method is not possible. The difference in the size of the traditional and the full text-based networks can be clearly seen.

PageRank (Page et al., 1999) is then calculated for each node for each network. We follow Chen et al. (2007) in their use of $\delta = 0.5$ for PageRank in scientific documents, from the assumption that readers of scientific papers are more likely to jump randomly to a new document compared to web surfers.

We emphasize that the experiments below are shown to provide comparisons between the traditional-based network construction methods and our proposed methods.

### 4.2 Finding prolific authors

Collaboration networks and citation networks can be used to find prolific authors (Chiang et al., 2013; Garfield, 2006). *Prolific authors* are authors who stand out based on their research output and contributions (Hasselback et al., 2003). We compare the results of the three different author-based networks by sorting the nodes of each network by their PageRank scores in descending order. We then calculate two metrics to measure author prolificity based on the information on Scopus[3] h-index, a widely used author-level metric and the quotient of the total citations over the number of documents the author has ($c/d$ metric). The second metric reflects prolificity more; an author is still influential if it has little documents with many citations. We then compute the average of the metrics of the first 10 authors for evaluation.

Table 2 shows the results of the experiments. It is shown clearly that the traditional co-occurrence network is inferior compared to the two full text-based networks in terms of the average h-index and the average $c/d$ metric. In terms of the average h-index, the full text-based citation network is the more superior network. This means that author citation graph is better in finding prolific authors if we need to also consider productivity. In terms of the average $c/d$ metric, the full text-based co-occurrence network is the more superior network. This means that the full text-based author collaboration network is better in finding prolific authors that emphasizes on the citation impact of the documents and does not consider productivity.

---

[3] https://www.scopus.com/search/submit/authorFreeLookup.uri

Table 2: Author collaboration and citation networks

| (a) traditional co-occurrence | | | (b) full text-based co-occurrence | | | (c) full text-based citation | | |
|---|---|---|---|---|---|---|---|---|
| Author | h | $c/d$ | Author | h | $c/d$ | Author | h | $c/d$ |
| HS MARK... | 76 | 62.45 | A JOUTEL | 41 | 92.47 | H CHABR... | 56 | 46.56 |
| TR BARR... | 29 | 38.09 | E TOURN... | 57 | 59.28 | A JOUTEL | 41 | 92.47 |
| AJ LAWR... | 39 | 21.49 | MG BOUS... | 87 | 58.19 | MG BOUS... | 87 | 58.19 |
| RG MORR... | 61 | 46.19 | H CHABR... | 56 | 46.56 | M DICHG... | 58 | 40.64 |
| M TRAYL... | 10 | 19.31 | K VAHEDI | 36 | 73.19 | E TOURN... | 57 | 59.28 |
| C LAMBE... | 8 | 14.38 | V DOMEN... | 16 | 162.88 | K VAHEDI | 36 | 73.19 |
| P BENJA... | 2 | 1.88 | MM RUCH... | 26 | 39.86 | HS MARK... | 76 | 62.45 |
| RL BROO... | 7 | 9.64 | J WEISS... | 112 | 154.07 | N PETERS | 24 | 34.43 |
| S BEVAN... | 22 | 40.41 | E MAREC... | 25 | 22.61 | F FAZEK... | 77 | 44.16 |
| B PATEL | 8 | 9.45 | EA CABA... | 23 | 13.64 | JM WARD... | 71 | 34.00 |
| **average** | *26.2* | *26.33* | **average** | **47.9** | **72.27** | **average** | **58.3** | 54.54 |

Table 3: Extracted biological entities per method

| traditional co-occurrence | **notch3**, **vascular dementia**, **stroke**, **hypertension**, alzheimer's disease, **migraine**, disease, vascular lesion, **ischemia**, **notch1**, multiple sclerosis, **amyloid angiopathy**, lacunar infarct, diabetes, single gene disorder, genetic disorder, **atherosclerosis**, allele, vascular, cortex |
|---|---|
| full text-based co-occurrence | **notch3**, **notch1**, **notch2**, **stroke**, alzheimer's disease, **hypertension**, multiple sclerosis, **vascular dementia**, dll4, **jag1**, **ischemic stroke**, **amyloid angiopathy**, **migraine**, disease, dll1, **fabry disease**, human disease, carasil, lacunar stroke, **atherosclerosis** |
| full text-based citation | **notch3**, **stroke**, **hypertension**, **caa**, alzheimer's disease, **notch1**, **migraine**, **atherosclerosis**, **vascular dementia**, lacunar infarct, disease, vascular lesion, **cvd**, diabetes, **notch2**, cortex, **ischemia**, dll4, skin, **brain atrophy** |

### 4.3 Finding important biological entities

We can also find important biological entities using co-occurrence and citation networks (Plake et al., 2006; Ding et al., 2013). We compare the results of the three different bio-entity-based networks by sorting the nodes of each network by their PageRank scores in descending order. We then remove all the other bio-entities and leave only the genes and diseases. For evaluation, we compare the first 20 bio-entities to MalaCards (Rappaport et al., 2013), a disease database that records related genes and diseases.

Table 3 shows the results of the experiments. The bold-faced entities are the important bio-entities. The traditional co-occurrence network provides the least number of important bio-entities with only nine entities found. Both the full text-based co-occurrence and the full text-based citation network found 12 important bio-entities. Interestingly, the co-occurrence network found one more gene (jag1) than the citation network.

### 4.4 Finding meaningful keywords

The keywords automatically extracted by the RAKE algorithm (Rose et al., 2010) may be general keywords and/or are not specific to our CADASIL dataset. The networks can be used to find the most meaningful keywords among the extracted keywords. We compare the results of the three different keyword-based networks by sorting the nodes of each network by their PageRank scores in descending order. For evaluation, we compare the first 20 keywords to MalaCards (Rappaport et al., 2013), which also contains other information regarding CADASIL.

Table 4: Extracted keywords per method

| traditional co-occurrence | homonymous visual field defect, **small vessel disease**, **vascular disease**, central retinal artery occlusion, intracranial pressure, optic disc edema, ischemic optic neuropathy, homonymous hemianopia, external carotid artery, ocular ischemic syndrome, **visual loss**, spontaneously, retinal ischemia, optic tract, retinal infarction, **cerebral white matter**, **central nervous system**, clinical presentation, cerebral atrophy, blood flow |
|---|---|
| full text-based co-occurrence | **cadasil**, **subcortical infarct**, **notch signaling**, risk factor, **vascular dementia**, **cognitive impairment**, **notch receptor**, **cerebral amyloid angiopathy**, **multiple sclerosis**, alagille syndrome, endothelial cell, **stroke**, **notch pathway**, **notch**, **alzheimer disease**, **cognitive decline**, risk, **notch signaling pathway**, disease, **small vessel disease** |
| full text-based citation | **notch signaling**, **cognitive impairment**, risk factor, endothelial cell, **cognitive decline**, **white matter**, risk, **alzheimer disease**, **notch receptor**, cognitive function, **cadasil**, cell, **stroke**, **subcortical infarct**, **ischemic stroke**, evidence, **notch**, **vascular risk factor**, previously, **notch signaling pathway** |

Table 5: Influential topics using PageRank.

| Topic 443 | Topic 297 | Topic 461 | Topic 243 | Topic 361 |
|---|---|---|---|---|
| risk | cell | study | matter | study |
| factor | notch | disease | disease | matter |
| diabetes | stem | research | svd | brain |
| hypertension | signaling | approach | lesion | impairment |
| smoking | differentiation | datum | wmh | association |
| disease | progenitor | treatment | stroke | lesion |
| stroke | fate | review | lacunar | mri |
| study | development | result | hyperintensity | volume |
| age | pathway | patient | vessel | wmh |
| mellitus | role | disorder | mri | wml |

Table 4 shows the results of the experiments. The bold-faced keywords are the meaningful extracted keywords. It is distinctly clear that the traditional-based method did not produce a lot of meaningful keywords, only extracting five. On the other hand, the full text-based co-occurrence network produced 14 meaningful keywords out of the 20 keywords extracted while the full text-based citation network produced 13 meaningful keywords out of the 20 keywords extracted.

## 4.5 Discovering influential research topics

Using the full text-based topic citation network, we can discover the top influential topics (Lee et al., 2016). Influential topics are topics that are frequently cited by other papers. In this paper, we present the influential topics in CADASIL research. Table 5 contains the top five influential topics based on PageRank. The most influential topic in CADASIL research is the research related to the cardiovascular disease (CVD) risk factors, such as high blood pressure, cholesterol, obesity, smoking, lack of physical ability and diabetes. The next most influential topic in CADASIL research is the research regarding notch signaling and how it regulates the differentiation of neural stem cells. The next three influential topics are case reports, research works on white matter hyperintensities (WMH) in small vessel diseases (SVD), and research works on cognitive impairment.

Table 6: Extracted genes per method

| out-degree citation (Ding et al., 2013) | traditional co-occurrence | full text-based co-occurrence | full text-based citation |
|---|---|---|---|
| insulin | oglcnac | **slc2a4** | **slc2a4** |
| large | p78 | gene | gene |
| impact | p180 | **sirt1** | **sirt1** |
| **lep** | p202 ptp1b gene | **nfe2l2** | **nfe2l2** |
| **tnf** | trem1 | met | ae |
| renin | **slc2a4** | glp1 ras | ppg |
| insulin receptor | dpp4 | ppg | met |
| set | **pparg** | **tp53** | **pten** |
| **mmp9** | sglt2 | ae | **tp53** |
| **mmp2** | ae | **pten** | sglt2 |

## 4.6 Metformin scarce data

In this section, we use Metformin as our data. Although Metformin is a widely research area in Medicine, we only use the first 1000 documents searched from the PubMed Central website to recreate a Metformin scarce data. We compare our methods to the traditional entity-entity citation network in Ding et al. (2013). They constructed the network using all the data found in the PMC website and used the abstracts of all the papers to extract the entities. Their results are then sorted using out-degree centrality. In this comparison, we use only the genes as the entities of our graph.

Table 6 shows the results of the experiments. The results in Ding et al. (2013) produced four related genes. It is clearly better compared to the traditional co-occurrence citation network with only two produced related genes. This is mainly because of the scarce data problem. However, both full text-based co-occurrence network and full text-based citation network produced five related genes, one more than the entity-entity citation network in Ding et al. (2013). This infers that even in the same setting with the same amount of data, the performance of the full text-based networks is comparable to or better than the performance of the traditional-based networks.

## 5 Conclusion

In this paper, we proposed an alternative method to constructing co-occurrence and citation networks. Instead of extracting entities from the given author and abstract metadata information, we proposed to look at the full text's reference section for the authors and the in-text citation context for the biological entities, keywords and topic. We especially recommend in using this to scarce data, where there is a lack in volume and in value. The advantages are three-fold: larger communities, clearer polarity, and citation emphasis.

We applied this method to research on CADASIL, a rare disorder. We constructed three co-occurrence networks (author, bio-entity, and keyword) and four citation networks (author, bio-entity, keyword, and topic) using the said method. We used it to different kinds of applications: finding prolific authors, finding important biological entities, finding meaningful keywords, and discovering influential topics. Compared to the traditional methods, full text-based methods perform noticeably better in finding significant entities. We also compared our method to the traditional-based entity-entity citation network in (Ding et al., 2013) and found out that even with the same quantity of data, the proposed full text-based network construction method is comparable to or better than the traditional-based network construction methods.

It is to note that looking at the full text instead of just the metadata information provides a more profound and defined analysis from the research articles. For future work, we can apply the methods and create a system to extract different kinds of entities from the full text and automatically construct the different kinds of networks given a set of research articles regarding a specific research area. This would further the research in biomedicine especially on rare diseases, genes, or chemicals.

## Acknowledgements

This project is supported fully by Microsoft Research.

## References

David M Blei, Andrew Y Ng, and Michael I Jordan. 2003. Latent dirichlet allocation. *Journal of machine Learning research*, 3(Jan):993–1022.

Michel Callon, Jean Pierre Courtial, and Francoise Laville. 1991. Co-word analysis as a tool for describing the network of interactions between basic and technological research: The case of polymer chemsitry. *Scientometrics*, 22(1):155–205.

Hugues Chabriat, Anne Joutel, Martin Dichgans, Elizabeth Tournier-Lasserve, and Marie-Germaine Bousser. 2009. Cadasil. *The Lancet Neurology*, 8(7):643–653.

Peng Chen, Huafeng Xie, Sergei Maslov, and Sidney Redner. 2007. Finding scientific gems with googles pagerank algorithm. *Journal of Informetrics*, 1(1):8–15.

Meng-Fen Chiang, Jiun-Jiue Liou, Jen-Liang Wang, Wen-Chih Peng, and Man-Kwan Shan. 2013. Exploring heterogeneous information networks and random walk with restart for academic search. *Knowledge and information systems*, 36(1):59–82.

Fabricio F Costa. 2014. Big data in biomedicine. *Drug discovery today*, 19(4):433–440.

Allan Peter Davis, Cynthia Grondin Murphy, Robin Johnson, Jean M Lay, Kelley Lennon-Hopkins, Cynthia Saraceni-Richards, Daniela Sciaky, Benjamin L King, Michael C Rosenstein, Thomas C Wiegers, et al. 2013. The comparative toxicogenomics database: update 2013. *Nucleic acids research*, 41(D1):D1104–D1114.

Ying Ding, Gobinda G Chowdhury, and Schubert Foo. 2001. Bibliometric cartography of information retrieval research by using co-word analysis. *Information processing & management*, 37(6):817–842.

Ying Ding, Min Song, Jia Han, Qi Yu, Erjia Yan, Lili Lin, and Tamy Chambers. 2013. Entitymetrics: Measuring the impact of entities. *PloS one*, 8(8):e71416.

Ying Ding. 2011. Scientific collaboration and endorsement: Network analysis of coauthorship and citation networks. *Journal of informetrics*, 5(1):187–203.

Eugene Garfield. 2006. The history and meaning of the journal impact factor. *Jama*, 295(1):90–93.

Jung Eun Hahm and Min Song. 2015. Detection of hidden knowledge using a citation-based approach based on swanson's abc model. *Journal of the Korean Society for information Management*, 32(2):87–103.

James R Hasselback, Alan Reinstein, and Edward S Schwan. 2003. Prolific authors of accounting literature. *Advances in Accounting*, 20:95–125.

Haiyan Hou, Hildrun Kretschmer, and Zeyuan Liu. 2008. The structure of scientific collaboration networks in scientometrics. *Scientometrics*, 75(2):189–202.

Doug Howe, Maria Costanzo, Petra Fey, Takashi Gojobori, Linda Hannick, Winston Hide, David P Hill, Renate Kania, Mary Schaeffer, Susan St Pierre, et al. 2008. Big data: The future of biocuration. *Nature*, 455(7209):47–50.

Yoo Kyung Jeong, Min Song, and Ying Ding. 2014. Content-based author co-citation analysis. *Journal of Informetrics*, 8(1):197–211.

Keeheon Lee, Hyojung Jung, and Min Song. 2016. Subject–method topic network analysis in communication studies. *Scientometrics*, pages 1–27.

Christopher D Manning, Mihai Surdeanu, John Bauer, Jenny Rose Finkel, Steven Bethard, and David McClosky. 2014. The stanford corenlp natural language processing toolkit. In *ACL (System Demonstrations)*, pages 55–60.

Ronald Margolis, Leslie Derr, Michelle Dunn, Michael Huerta, Jennie Larkin, Jerry Sheehan, Mark Guyer, and Eric D Green. 2014. The national institutes of health's big data to knowledge (bd2k) initiative: capitalizing on biomedical big data. *Journal of the American Medical Informatics Association*, 21(6):957–958.

Vivien Marx. 2013. Biology: The big challenges of big data. *Nature*, 498(7453):255–260.

Mark EJ Newman. 2001. Scientific collaboration networks. i. network construction and fundamental results. *Physical review E*, 64(1):016131.

Mark EJ Newman. 2004. Coauthorship networks and patterns of scientific collaboration. *Proceedings of the national academy of sciences*, 101(suppl 1):5200–5205.

Lawrence Page, Sergey Brin, Rajeev Motwani, and Terry Winograd. 1999. The pagerank citation ranking: bringing order to the web.

Karen E Pettigrew and Lynne EF McKechnie. 2001. The use of theory in information science research. *Journal of the American Society for Information Science and Technology*, 52(1):62–73.

Conrad Plake, Torsten Schiemann, Marcus Pankalla, Jörg Hakenberg, and Ulf Leser. 2006. Alibaba: Pubmed as a graph. *Bioinformatics*, 22(19):2444–2445.

Noa Rappaport, Noam Nativ, Gil Stelzer, Michal Twik, Yaron Guan-Golan, Tsippi Iny Stein, Iris Bahir, Frida Belinky, C Paul Morrey, Marilyn Safran, et al. 2013. Malacards: an integrated compendium for diseases and their annotation. *Database*, 2013:bat018.

Thomas C Rindflesch, Lorraine Tanabe, John N Weinstein, and Lawrence Hunter. 2000. Edgar: extraction of drugs, genes and relations from the biomedical literature. In *Pacific Symposium on Biocomputing. Pacific Symposium on Biocomputing*, page 517. NIH Public Access.

Stuart Rose, Dave Engel, Nick Cramer, and Wendy Cowley. 2010. Automatic keyword extraction from individual documents. *Text Mining*, pages 1–20.

Gerard Salton, Anita Wong, and Chung-Shu Yang. 1975. A vector space model for automatic indexing. *Communications of the ACM*, 18(11):613–620.

Burr Settles. 2005. Abner: an open source tool for automatically tagging genes, proteins and other entity names in text. *Bioinformatics*, 21(14):3191–3192.

Min Song, SuYeon Kim, Guo Zhang, Ying Ding, and Tamy Chambers. 2014. Productivity and influence in bioinformatics: A bibliometric analysis using pubmed central. *Journal of the Association for Information Science and Technology*, 65(2):352–371.

Min Song, Won Chul Kim, Dahee Lee, Go Eun Heo, and Keun Young Kang. 2015. Pkde4j: Entity and relation extraction for public knowledge discovery. *Journal of biomedical informatics*, 57:320–332.

Teja Tscharntke, Michael E Hochberg, Tatyana A Rand, Vincent H Resh, and Jochen Krauss. 2007. Author sequence and credit for contributions in multiauthored publications. *PLoS Biol*, 5(1):e18.

Erjia Yan, Ying Ding, Blaise Cronin, and Loet Leydesdorff. 2013. A bird's-eye view of scientific trading: Dependency relations among fields of science. *Journal of Informetrics*, 7(2):249–264.

Xiaoshi Yin, Jimmy Xiangji Huang, and Zhoujun Li. 2011. Mining and modeling linkage information from citation context for improving biomedical literature retrieval. *Information processing & management*, 47(1):53–67.

Karol Życzkowski. 2010. Citation graph, weighted impact factors and performance indices. *Scientometrics*, 85(1):301–315.

# Named Entity Recognition in Swedish Health Records
## with Character-Based Deep Bidirectional LSTMs


**Simon Almgren[1], Sean Pavlov[1], Olof Mogren**[*]
Chalmers University of Technology, Sweden
*olof@mogren.one



## Abstract

We propose an approach for named entity recognition in medical data, using a character-based
deep bidirectional recurrent neural network. Such models can learn features and patterns based
on the character sequence, and are not limited to a fixed vocabulary. This makes them very well
suited for the NER task in the medical domain. Our experimental evaluation shows promising
results, with a 60% improvement in $F_1$ score over the baseline, and our system generalizes well
between different datasets.


## 1   Introduction

Named Entity Recognition (NER) is the task of finding mentions of named entities in a text. In nonmedical NER, entity classes are typically people, organizations, and locations. It is one of the fundamental Natural Language Processing (NLP) tasks and has been studied extensively.

In this paper, we approach the problem of finding medical entities such as (1) *disorders and findings*, (2) *pharmaceutical drugs*, and (3) *body structure*. Our proposed
method uses deep bidirectional character-based recurrent
neural networks (RNNs), trained in an end-to-end fashion to perform both boundary detection and classification
at the same time.

There are a number of properties that make this problem especially challenging in biomedical text (Zhou et
al., 2004). Firstly, names composed of multiple words
are frequently used to describe an entity, highlighting the
requirement of good boundary detection on an NER system. Secondly, one noun can be part of a mention of

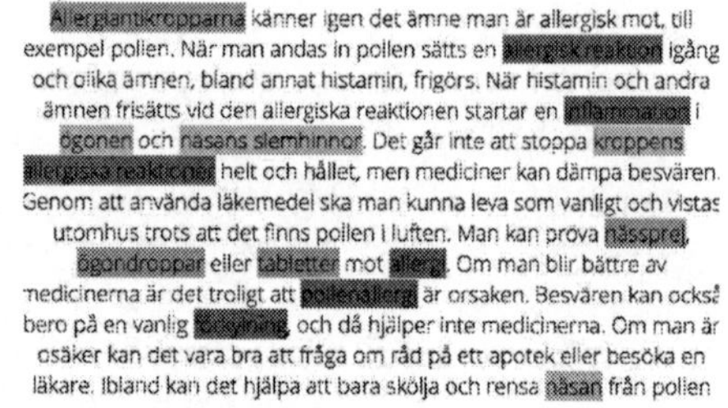

Figure 1: A Swedish medical example text
with NER tags illustrated with colour.

several entities at the same time E.g: "91 and 84 kDa proteins" consists of two entity names: "91 kDa
proteins" and "84 kDa proteins". Thirdly, it is common to write the same biomedical entity in different
ways, e.g: "N-acetylcysteine", "N-acetyl-cysteine", "NAcetylCysteine". Lastly, ambiguous mentions
are common, including abbreviations that refer to different things in different contexts. (The examples
above are from Zhou et al. (2004)).

Our proposed method has a number of benefits over previous work: Firstly, the model can simultaneously recognize and classify entity mentions. Secondly, using an end-to-end neural network approach
eliminates the need for feature engineering. All features needed are learned by the model during training. Thirdly, because our model works on the raw character sequence, it does not suffer from out-of-
vocabulary terms, it can learn that different character patterns represent the same thing, and it can learn
the typical character-based features often used in traditional machine learning based solutions to NER.

We evaluate the model on Swedish health records in the Stockholm EPR corpus and obtain promising
results. We also note that the method generalizes well between different datasets.

---

1. Equal contribution.






*Proceedings of the Fifth Workshop on Building and Evaluating Resources for Biomedical Text Mining (BioTxtM 2016),*
pages 30–39, Osaka, Japan, December 12th 2016.

## 2   Background

A recurrent neural network (RNN) is a feedforward artificial neural network that can model a sequence of arbitrary length, using weight sharing between each position in the sequence. In a language setting, it is common to model sequences of words, in which case each input $x_t$ is the vector representation of a word. In the basic RNN variant, the transition function is a linear transformation of the hidden state and the input, followed by a pointwise nonlinearity:

$$h_t = \tanh(Wx_t + Uh_{t-1} + b),$$

where $W$ and $U$ are trainable weight matrices, $b$ is a bias term, and tanh is the nonlinearity.

Basic RNNs struggle with learning long dependencies and suffer from the vanishing gradient problem. This makes RNN models difficult to train (Hochreiter, 1998; Bengio et al., 1994), and provoked the development of the Long Short Term Memory (LSTM) (Schmidhuber and Hochreiter, 1997), that to some extent solves these shortcomings. An LSTM is an RNN where the cell at each step $t$ contains an internal memory vector $c_t$, and three gates controlling what parts of the internal memory will be kept (the forget gate $f_t$), what parts of the input that will be stored in the internal memory (the input gate $i_t$), as well as what will be included in the output (the output gate $o_t$). In essence, this means that the following expressions are evaluated at each step in the sequence, to compute the new internal memory $c_t$ and the cell output $h_t$. Here "$\odot$" represents element-wise multiplication.

$$
\begin{aligned}
i_t &= \sigma(W^{(i)}x_t + U^{(i)}h_{t-1} + b^{(i)}), \\
f_t &= \sigma(W^{(f)}x_t + U^{(f)}h_{t-1} + b^{(f)}), \\
o_t &= \sigma(W^{(o)}x_t + U^{(o)}h_{t-1} + b^{(o)}), \\
u_t &= \tanh(W^{(u)}x_t + U^{(u)}h_{t-1} + b^{(u)}), \\
c_t &= i_t \odot u_t + f_t \odot c_{t-1}, \\
h_t &= o_t \odot \tanh(c_t).
\end{aligned}
\tag{1}
$$

Most RNN based models work on word level. Words are coded as a one-hot vector, and each word is associated with an internally learned embedding vector. In this work, we propose a character-level model that is able to learn features based on arbitrary parts of the character sequence.

LSTM networks have been used successfully for language modelling, sentiment analysis (Tang et al., 2015), textual entailment (Rocktäschel et al., 2016), and machine translation (Sutskever et al., 2014). In the following sections, we will see that the learned features are also suitable for recognizing and classifying mentions of medical entities in health record data.

## 3   Named Entity Recognition with Character-Based Deep Bidirectional LSTMs

In this paper, we propose a character based RNN model with deep bidirectional LSTM cells (BiLSTM) to do Named Entity Recognition in the medical domain (see Figure 2). The model is trained and evaluated on medical texts in Swedish. It has a softmax output layer with four outputs corresponding to each position in the input sequence, representing the three different entity labels, and a special label for all non-entity characters.

The model is trained end-to-end using backpropagation and the Adam optimizer (Diederik Kingma, 2015) to perform entity classification on a character-by-character basis. A neural network learns to internally represent data with representations that are useful for the task. This is an effect of using backpropagation, and allows us to eliminate all manual feature engineering, enabling quick deployment of our system.

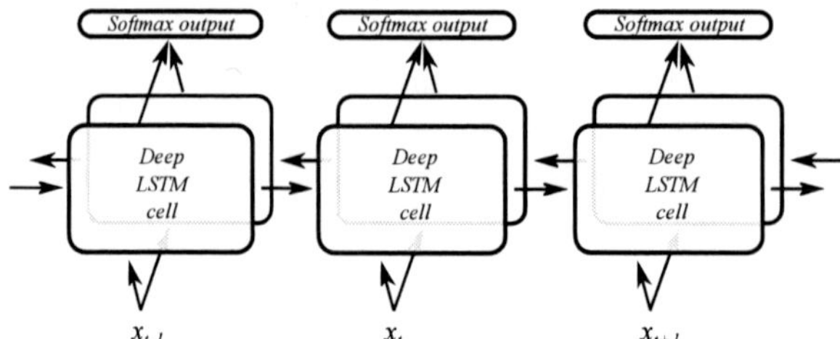


Figure 2: A deep bidirectional LSTM network. At each input $x_t$, the model is trained to output a prediction $y_t$ of the correct entity class. In this paper, each block is a deep LSTM cell (see Figure 3), and the network is trained using backpropagation through time (BPTT).

### 3.1 Character classification

Our model works on the raw character sequence of the input document. This is an approach that has proven to work well in some other NLP applications, (see e.g. Luong and Manning (2016), **?**)).

Compared to a word-based sequence model, this means that we can use a much smaller vocabulary for the input tokens. Traditional (non-neural) entity recognition systems typically rely heavily on hand-engineered character-based features, such as capitalization, numerical characters, word prefixes and suffixes (Ratinov and Roth, 2009). Having the capacity of learning this kind of features automatically is what motivated us to use this kind of model. A character-based model does not rely on words being in its vocabulary: any word can be represented, as long as it is written with the given alphabet.

The character sequence model computes one classification output per input character. The label is one of: (1) *disorders and findings*, (2) *pharmaceutical drugs*, (3) *body structure*, (4) *non-entity term*. Using these labels (including the special "non-entity" label), we can simultaneously recognize and classify entity mentions by computing one label per character in

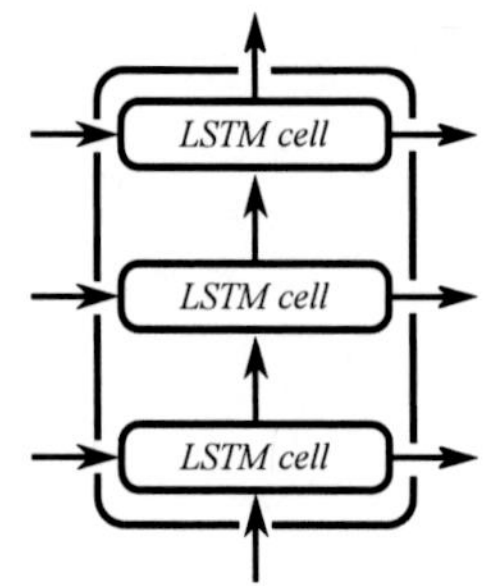


Figure 3: A deep LSTM cell, consisting of 3 internally stacked LSTM cells.

the input text. This means that we can interpret each connected subsequence with the same classification as an entity mention.

However, there are some special cases: Firstly, to handle the situation when sporadic characters are classified inconsistently, we treat the character classifications as a voting mechanism for each word, and the majority class is chosen. Secondly, if a space between two tokens is classified consistently with the two tokens, both tokens are interpreted as belonging to the same entity mention. If the space is classified as a non-entity character, the two tokens are treated as two different entity mentions.

## 4 Experimental setup

This section explains the set-up of the empirical study of our model.

### 4.1 Model layout

We used a deep bidirectional recurrent neural network with LSTM cells. The depth of the LSTM cells was set to 3, and we used 128 hidden units in the LSTM cells. The model was implemented using Tensorflow. Learning rate: 0.002, decay rate: 0.975. Using drop-out on activations from the input embedding layers as well as on the LSTM output activations were evaluated, but was left out in the final version. See Section 4.4 for details on hyperparameters. The source code of our model is available on Github[1].

---

[1] `https://github.com/withtwist/medical-ner/`

## 4.2 Seed-term collection

Seed-terms are used both to build the datasets (see Section 4.3), and to build up the representations for the classification centroids in the BOW baseline method. Seed-terms were extracted from two taxonomies, SweMeSH[2], a taxonomy of Swedish medical terms and Snomed CT[3], consisting of Swedish medical concept terms. Using the hierarchical structure of the two taxonomies, all terms that was descendants of each of our predefined categories was extracted and considered seed-terms. The following predefined categories was used for the extraction: *disorder & finding* (*sjukdom & symtom*), *pharmaceutical drug* (*läkemedel*) and *body structure* (*kroppsdel*). The choice of these main entity classes was aligned with Skeppstedt et al. (2014).

## 4.3 Datasets

We use an approach similar to Mintz et al. (2009) to obtain the data needed for training and evaluation. The datasets that we prepared for training, validating and testing our model are available for download at `https://github.com/olofmogren/biomedical-ner-data-swedish/`.

The *Läkartidningen* corpus was originally presented by Kokkinakis and Gerdin (2010), and contains articles from the Swedish journal for medical professionals. This was annotated for NER as a part of this work. All occurrences of seed-terms were extracted (see Section 4.2), along with a context window of 60 characters (approximately ten words). The window is positioned so that the entity mention is located randomly within the sequence. In addition, negative training examples were extracted in order to prevent the model from learning that classified entities always occur in every sequence. All the characters in these negative training examples had the same "non-entity" label. Neural models typically benefit from large amounts of training data. To increase the amount of training data, each occurrence of seed-terms were extracted three more times, where the window was shifted by a random number of steps. The resulting data is a total of 775,000 of sequences with 60 characters each. 10% of the data is negative data, where every character has the "non-entity" label.

Another dataset was built from medical articles on the *Swedish Wikipedia*. Firstly, an initial list of medical domain articles were chosen manually and fetched. Secondly, articles were fetched that were linked from the initial articles. Finally, the seed-terms list (see Section 4.2) was used to create the labels and extract training examples of 60 character sequences, in the same way as was done with *Läkartidningen*.

*1177 Vårdguiden* iis a web site provided by the Swedish public health care authorities, containing information, counselling, and other health-care services. The corpus consists of 15 annotated documents downloaded during May 2016. This dataset was manually annotated with the seed-terms list as support (see Section 4.2). The resulting dataset has 2740 annotations, out of which 1574 are *disorder and finding*, 546 are *pharmaceutical drug*, and 620 are *body structure*.

The *Stockholm Electronic Patient Record (EPR) Clinical Entity Corpus* (Dalianis et al., 2012) is a dataset with health records of over two million patients at Karolinska University Hospital in Stockholm encompassing the years 2006-2014. It consists of 7946 documents containing real-world anonymized health records with annotations in 4 categories: *disorder, finding, drug* and *body structure*. Since we have a category where "*disorder*" and "*finding*" are bundled together they were considered the same.

*Läkartidningen*, *Swedish Wikipedia*, and *1177 Vårdguiden* are all datasets with rather high quality text, most of it even professionally edited. This is in stark contrast to the text in *Stockholm EPR* where misspellings are common, there are reduntant parts in many records, and writing style is highly diverse (Dalianis et al., 2009).

---

[2]`http://mesh.kib.ki.se/swemesh/swemesh_se.cfm`
[3]`http://www.socialstyrelsen.se/nationellehalsa/snomed-ct`

## 4.4 Hyperparameter search

A number of settings for hyperparameters were explored
during development. In the variations listed below, one hyperparameter at the time is varied and evaluated, and if we
saw an improvement, the change of setting was retained.
A more thorough hyperparameter investigatin is left for future work. For the three first experiments, dropout was
used on the activations from the embedding layer, as well
as on the activations on the LSTM outputs. (See Section 4.1 for details).

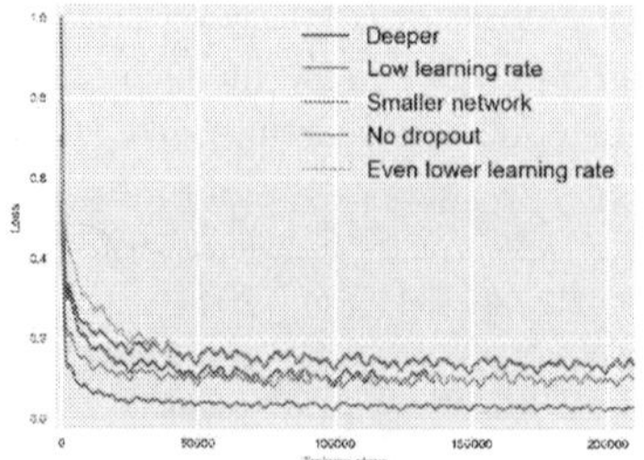

Figure 4: Validation loss.

*Deeper*: A model using 4 stacked LSTM cells. Learning
rate: 0.05, decay rate: 0.975, drop probability: 0.5. *Low
learning rate*: LSTK depth: 3, learning rate: 0.002, decay
rate: 0.975. Lowering the learning rate proved useful and 0.002 became the default setting for learning
rate. Drop probability: 0.5. *Smaller network*: 64 hidden units in each LSTM cell. LSTM depth: 3:
learning rate: 0.002, decay rate: 0.975, drop probability: 0.5. *No dropout*: This model left all the settings
as default but removed dropout entirely. 128 hidden units in each LSTM cell, LSTM depth: 3, learning
rate: 0.002 and decay rate: 0.975. This setting proved to be the best, which meant that the default settings
subsequently never used dropout. *Even lower learning rate*: Learning rate: 0.0002 and decay rate: 0.975.
No drop-out.

See Figure 4 for the validation performance of the different settings. The resulting model used in
the final experiments reported in Section 5 had 3 stacked LSTM layers with 128 hidden units in each.
Learning rate: 0.002, decay rate 0.975, and no drop-out.

## 4.5 Baselines

Two baselines were implemented and used. The **dictionary baseline** simply consist of dictionary look-
ups of the encountered words in the list of seed-terms. The **BOW (Bag-Of-Words) baseline** is based
on Zhang and Elhadad (2013). The original version was developed for and evaluated on medical texts
in English. The approach considers each noun-phrase as an entity candidate, and represents each candidate using a weighted bag-of-words-vector for the context. The required preprocessing is tokenization,
sentence splitting, part-of-speech-tagging, and noun phrase chunking. The first steps was done using
GATE (Cunningham et al., 2011) and OpenNLP[4], while noun phrase chunking was done using Swe-
SPARK (Aycock, 1998). An IDF threshold is used to filter out uncommon or unspecific noun phrases.
Then for each category the algorithm builds an average context vector representing the mentions in a
training corpus. We used a triangular window for the context vectors, giving the central words a weight
of 20, and context words the weight of $1/k$, where $k$ is the distance from the central word. Mentions with
a cosine similarity lower than 0.005 to any of the category vectors was discarded. Candidate mentions
that have a difference between the top two scoring categories that is lower than 0.7 are also discarded.

Zhang and Elhadad (2013) used one bag-of-words vector for the internal words of an entity mention,
and one for the context words of the mention. The two vectors were then concatenated, resulting in a
vector which is twice the size of their vocabulary. Since the bag-of-words-vectors are already sparse to
begin with, we added them together instead and made it possible to use a larger vocabulary size.

## 4.6 Training

Development and training were performed using text from *Läkartidningen* (Kokkinakis and Gerdin,
2010). Validation was done using the *Medical Wikipedia* dataset. Training was done using the Adam
optimizer (Diederik Kingma, 2015).

---

[4] http://opennlp.sourceforge.net/models-1.5/

### 4.7 Evaluation

Evaluation of the proposed model was performed on two different datasets: *Stockholm EPR corpus* (Dalianis et al., 2012), with anonymized health record data in Swedish, and *1177 Vårdguiden*.

We report $F_1$ scores for total named entity recognition, as well as only entity classification (given correct boundary detection, we report scores of the entity classification performed by the system).

In the BOW baseline the entities are determined before hand while the Char-BiLSTM model recognizes and classifies them as it traverses the document.

## 5 Results

This section presents the results of the experimental evaluation of the proposed model. Table 1 shows the results of running the dictionary baseline model on *Stockholm EPR corpus* (Dalianis et al., 2012). The baseline model achieves a precision of over 0.70 on *disorder & Finding* and *body structure*, but is substantially lower for *pharmaceutical drug*. It has a higher precision than recall in general due to the fact that if a match is found it is probably correct. The algorithm got a precision of 0.67, a recall of 0.12 and an $F_1$ score of 0.20.

| Category | P | R | $F_1$ |
|---|---|---|---|
| Disorder & finding | 0.76 | 0.12 | 0.20 |
| Pharmaceutical drug | 0.25 | 0.04 | 0.07 |
| Body structure | 0.70 | 0.29 | 0.41 |
| **Total** | 0.67 | 0.12 | 0.20 |

Table 1: Dictionary baseline performance on the *Stockholm EPR corpus*. Although total precision is reasonably good (0.67), the precision (0.12) is not.

The evaluation of the Char-BiLSTM model was performed on 733 real-world examples of health records from *Stockholm EPR corpus* (Dalianis et al., 2012). Since the data is very sensitive the evaluation was not performed by ourselves but instead the holder of the data performed the evaluations.

**Char-BiLSTM overall results**: The results in Table 2 shows the results of the Char-BiLSTM model. Both *disorder & finding* and *body structure* have a much higher precision than recall, while *pharmaceutical drug* is better balanced.

| Category | P | R | $F_1$ |
|---|---|---|---|
| Disorder & findings | 0.72 | 0.18 | 0.29 |
| Pharmaceutical drugs | 0.69 | 0.43 | 0.53 |
| Body structure | 0.46 | 0.28 | 0.35 |
| **Total** | 0.67 | 0.24 | 0.35 |

Table 2: Results for Char-BiLSTM on *Stockholm EPR corpus*. The model obtains a total precision that matches the dictionary baseline (0.67), and a recall that is much higher than the baseline (0.24).

**Char-BiLSTM results, classification only**: Given the boundaries for the entities in *Stockholm EPR corpus*, the performance of the Char-BiLSTM model (performing only classification of the given entities) is given in Table 3. The table shows promising results for both the category *disorder & finding* and *pharmaceutical drug* which has an $F_1$ score of 0.81 and 0.74 respectively. *body structure* shows a weaker $F_1$ score of 0.47. The model got an overall $F_1$ score of 0.75.

We compare our system with the two baselines using the *1177 Vårdguiden corpus*. Since the BOW baseline detects boundaries using whole noun phrases, we re-ran the experiments, adjusting the evaluation data, so that the boundaries were complete noun-phrases.

| Category | P | R | $F_1$ |
|---|---|---|---|
| Disorder & findings | 0.92 | 0.73 | 0.81 |
| Pharmaceutical drugs | 0.64 | 0.87 | 0.74 |
| Body structure | 0.36 | 0.68 | 0.47 |
| **Total** | 0.75 | 0.75 | 0.75 |

Table 3: Entity classification results of Char-BiLSTM on *Stockholm EPR corpus*. Given the entity boundaries, we can see that the classification of entities work very well, obtaining a total $F_1$ score of 0.75.

Comparing the results of the models in Table 4, we see that the BOW baseline does not perform well due to the wider boundaries that it detects. This can clearly be seen in the experiments with adjusted data, where it performs around 47% better. All models have around 0.50 in precision, except for the adjusted BOW baseline which comes in at 0.32. Recall is much lower (between 0.08 and 0.17), except for the BiLSTM model which has a recall of 0.21. This is the main reason why the BiLSTM model has the highest $F_1$ score with 0.29. At the second place comes the adjusted BOW baseline at 0.22, followed by the dictionary baseline model at 0.22 and lastly the BOW baseline with 0.15. Even though the dictionary baseline model and the adjusted BOW baseline have similar performance scores, we can see that their precision and recall are vastly different. The dictionary baseline model has a high precision and a low recall, while the adjusted BOW baseline is more balanced between precision and recall.

| Model | P | R | $F_1$ |
|---|---|---|---|
| Dictionary | 0.54 | 0.14 | 0.22 |
| BOW | **0.55** | 0.09 | 0.15 |
| BOW (adj.) | 0.32 | 0.17 | 0.22 |
| Char-BiLSTM | 0.48 | **0.21** | **0.29** |

Table 4: Comparison of the results between each model on *1177 Vårdguiden*.

## 6   Related work

Supervised NER has been thoroughly explored in the past. Finkel et al. (2005) used Conditional Random Fields (CRF), a technique often used for NER. Zhou et al. (2004) used Hidden Markov Models (HMMs) along with extensive feature engineering to perform NER on medical texts. State-of-the-art in the medical domain have been achieved by Wang and Patrick (2009) with a combination of CRF, Support Vector Machines (SVM) and Maximum Entropy (ME) to recognize and classify entities.

Skeppstedt et al. (2014) currently holds the state-of-the-art in Swedish for the medical domain, based on CRF.

Recently, a number of papers have proposed using RNNs for sequence labelling tasks. Cícero Nogueira dos Santos (2015) presented a model that learns word embeddings along with character embeddings from a convolutional layer, which are used in a window-based fixed feed forward neural network. Huang et al. (2015) proposed a bidirectional LSTM model, but it used traditional feature engineering, and the classification was performed using a CRF layer in the network. In contrast, our proposed model learns all its features, and can be trained efficiently with simple backpropagation and stochastic gradient descent. Ma and Hovy (2016) presented a model that uses a convolutional network to compute representations for parts of words. Then the representations are combined with some character-level features and fed into a bidirectional LSTM network, and finally a CRF performs the labelling. Chiu and Nichols (2016) presented a similar model but with a softmax output instead of the CRF layer. Like our system, the models are trained end-to-end and obtains good results on standard NER evaluations, however our system is conceptually simpler, and learns all of its features directly from the character stream. Lample et al. (2016) presented two different architechtures, one using LSTMs and CRFs, and one using a shift-reduce approach. Gillick et al. (2016) presented a character-based model with LSTM units similar to a translation model, but instead of decoding into a different language, the state from the encoder is decoded into a sequence of tags.

Learning representations for text is important for many other tasks within natural language processing. A common way of representing sequences of words is to use some form of word embeddings (Mikolov et al., 2013; Pennington et al., 2014; Collobert and Weston, 2008), and for each word in the sequence, do an element-wise addition (Mitchell and Lapata, 2010). This approach works well for many applications, such as phrase similarity, multi-document summarization (Mogren et al., 2015), and word sense induction (Kågebäck et al., 2015), even though it disregards the order of the words. In contrast, RNNs and LSTMs (Hochreiter and Schmidhuber, 1997) learn representations for text that takes the order into account. They have been successfully been applied to sentiment analysis (Tang et al., 2015), question answering systems (Hagstedt P Suorra and Mogren, 2016), and machine translation (Sutskever et al., 2014).

Character-based neural sequence models have recently been presented to tackle the problem of out-of-vocabulary terms in neural machine translation (Luong and Manning, 2016; Chung et al., 2016) and for language modelling (Kim et al., 2016).

## 7   Discussion

The results of the empirical evaluation of the proposed system show some interesting points, suggesting that this approach should be researched further.

We have evaluated our model on the Stockholm EPR corpus of Swedish health records, but we did not compare our scores with other approaches that was evaluated on the same dataset. The reason is that we were unable to do a fair comparison since our model was trained on other data. We believe that our scores are competitive, and indicates that the model is promising. While systems that were trained on data from the same corpus show better performance in the evaluation on *Stockholm EPR* (Skeppstedt et al., 2014), we note that our solution can be trained on a dataset that is quite diferent from the test set. This can be explained in part with the documented robustness of character-based recurrent neural models to misspellings and out-of-vocabulary terms.

We are convinced that our solution would be able to obtain even better scores if able to train on the same data.

## 8   Conclusions

In this paper, we have proposed a character-based deep bidirectional LSTM model (Char-BiLSTM) to perform named entity recognition in Swedish health records. We beat two different baseline methods on the task, and show that this is a promising research direction. The proposed model obtains an $F_1$ score of 0.35 which is about 60% better than the BOW baseline (Zhang and Elhadad, 2013). Our model learns all the features it needs, and therefore eliminates the need for feature engineering. We have seen that a character-based neural model adapts well to this domain, and in fact that it is able to generalize from relatively well-written training data to test-data with lesser quality text.

## Acknowledgments

This work has been done within the project "Data-driven secure business intelligence", grant IIS11-0089 from the Swedish Foundation for Strategic Research (SSF).

## References

John Aycock. 1998. Compiling little languages in python. In *Proceedings of the 7th International Python Conference*, pages 69–77.

Yoshua Bengio, Patrice Simard, and Paolo Frasconi. 1994. Learning long-term dependencies with gradient descent is difficult. *Neural Networks, IEEE Transactions on*, 5(2):157–166.

Jason Chiu and Eric Nichols. 2016. Named entity recognition with bidirectional lstm-cnns. *Transactions of the Association for Computational Linguistics*, 4:357–370.

Junyoung Chung, Kyunghyun Cho, and Yoshua Bengio. 2016. A character-level decoder without explicit segmentation for neural machine translation. In *Proceedings of the 54th Annual Meeting of the Association for Computational Linguistics (Volume 1: Long Papers)*, pages 1693–1703, Berlin, Germany, August. Association for Computational Linguistics.

Victor Guimarães Cícero Nogueira dos Santos. 2015. Boosting named entity recognition with neural character embeddings. *Proceedings of NEWS 2015 The Fifth Named Entities Workshop*.

Ronan Collobert and Jason Weston. 2008. A unified architecture for natural language processing: Deep neural networks with multitask learning. In *Proceedings of the 25th international conference on Machine learning*, pages 160–167. ACM.

Hamish Cunningham, Diana Maynard, Kalina Bontcheva, Valentin Tablan, Niraj Aswani, Ian Roberts, Genevieve Gorrell, Adam Funk, Angus Roberts, Danica Damljanovic, Thomas Heitz, Mark A. Greenwood, Horacio Saggion, Johann Petrak, Yaoyong Li, and Wim Peters. 2011. *Text Processing with GATE (Version 6)*.

Hercules Dalianis, Martin Hassel, and Sumithra Velupillai. 2009. The stockholm epr corpus—characteristics and some initial findings. In *Proceedings of the 14th International Symposium on Health Information Management Research - iSHIMR*.

Hercules Dalianis, Martin Hassel, Aron Henriksson, and Maria Skeppstedt. 2012. Stockholm epr corpus: A clinical database used to improve health care. *Swedish Language Technology Conference*, pages 17–18.

Jimmy Ba Diederik Kingma. 2015. Adam: A method for stochastic optimization. *International Conference on Learning Representations*.

Jenny Rose Finkel, Trond Grenager, and Christopher Manning. 2005. Incorporating non-local information into information extraction systems by gibbs sampling. In *Proceedings of the 43rd Annual Meeting on Association for Computational Linguistics*, pages 363–370. Association for Computational Linguistics.

Dan Gillick, Cliff Brunk, Oriol Vinyals, and Amarnag Subramanya. 2016. Multilingual language processing from bytes. In *Proceedings of the 2016 Conference of the North American Chapter of the Association for Computational Linguistics: Human Language Technologies*, pages 1296–1306, San Diego, California, June. Association for Computational Linguistics.

Jacob Hagstedt P Suorra and Olof Mogren. 2016. Assisting discussion forum users using deep recurrent neural networks. *Proceedings of the 1st Workshop on Representation Learning for NLP at ACL 2016*, page 53.

Sepp Hochreiter and Jürgen Schmidhuber. 1997. Long short-term memory. *Neural computation*, 9(8):1735–1780.

Sepp Hochreiter. 1998. The vanishing gradient problem during learning recurrent neural nets and problem solutions. *International Journal of Uncertainty, Fuzziness and Knowledge-Based Systems*, 6(02):107–116.

Zhiheng Huang, Wei Xu, and Kai Yu. 2015. Bidirectional lstm-crf models for sequence tagging. *arXiv preprint arXiv:1508.01991*.

Mikael Kågebäck, Fredrik Johansson, Richard Johansson, and Devdatt Dubhashi. 2015. Neural context embeddings for automatic discovery of word senses. In *Proceedings of NAACL-HLT*, pages 25–32.

Yoon Kim, Yacine Jernite, David Sontag, and Alexander M Rush. 2016. Character-aware neural language models. In *Thirtieth AAAI Conference on Artificial Intelligence*.

Dimitrios Kokkinakis and Ulla Gerdin. 2010. Läkartidningens arkiv i en ny skepnad - en resurs för forskare, läkare och allmänhet. *Språkbruk*, pages 22–28.

Guillaume Lample, Miguel Ballesteros, Sandeep Subramanian, Kazuya Kawakami, and Chris Dyer. 2016. Neural architectures for named entity recognition. In *Proceedings of the 2016 Conference of the North American Chapter of the Association for Computational Linguistics: Human Language Technologies*, pages 260–270, San Diego, California, June. Association for Computational Linguistics.

Minh-Thang Luong and Christopher D. Manning. 2016. Achieving open vocabulary neural machine translation with hybrid word-character models. In *Proceedings of the 54th Annual Meeting of the Association for Computational Linguistics (Volume 1: Long Papers)*, pages 1054–1063, Berlin, Germany, August. Association for Computational Linguistics.

Xuezhe Ma and Eduard Hovy. 2016. End-to-end sequence labeling via bi-directional lstm-cnns-crf. In *Proceedings of the 54th Annual Meeting of the Association for Computational Linguistics (Volume 1: Long Papers)*, pages 1064–1074, Berlin, Germany, August. Association for Computational Linguistics.

Tomas Mikolov, Kai Chen, Greg Corrado, and Jeffrey Dean. 2013. Efficient estimation of word representations in vector space. In *ICLR*.

Mike Mintz, Steven Bills, Rion Snow, and Dan Jurafsky. 2009. Distant supervision for relation extraction without labeled data. In *Proceedings of the Joint Conference of the 47th Annual Meeting of the ACL and the 4th International Joint Conference on Natural Language Processing of the AFNLP: Volume 2-Volume 2*, pages 1003–1011. Association for Computational Linguistics.

Jeff Mitchell and Mirella Lapata. 2010. Composition in distributional models of semantics. *Cognitive science*, 34(8):1388–1429.

Olof Mogren, Mikael Kågebäck, and Devdatt Dubhashi. 2015. Extractive summarization by aggregating multiple similarities. In *Recent Advances in Natural Language Processing*, page 451.

Jeffrey Pennington, Richard Socher, and Christopher D Manning. 2014. Glove: Global vectors for word representation. In *EMNLP*, volume 14, pages 1532–43.

Lev Ratinov and Dan Roth. 2009. Design challenges and misconceptions in named entity recognition. In *Proceedings of the Thirteenth Conference on Computational Natural Language Learning*, pages 147–155. Association for Computational Linguistics.

Tim Rocktäschel, Edward Grefenstette, Karl Moritz Hermann, Tomáš Kočiský, and Phil Blunsom. 2016. Reasoning about entailment with neural attention. In *International Conference on Learning Representations*.

Jürgen Schmidhuber and Sepp Hochreiter. 1997. Long short-term memory. *Neural computation*, 7(8):1735–1780.

Maria Skeppstedt, Maria Kvist, H. Nilsson Gunnar, and Dalianis Hercules. 2014. Automatic recognition of disorders, findings, pharmaceuticals and body structures from clinical text: An annotation and machine learning study. *Journal of Biomedical Informatics*, 49:148–158.

Ilya Sutskever, Oriol Vinyals, and Quoc V Le. 2014. Sequence to sequence learning with neural networks. In *Advances in neural information processing systems*, pages 3104–3112.

Duyu Tang, Bing Qin, and Ting Liu. 2015. Document modeling with gated recurrent neural network for sentiment classification. In *Proceedings of the 2015 Conference on Empirical Methods in Natural Language Processing*, pages 1422–1432.

Yefeng Wang and Jon Patrick. 2009. Cascading classifiers for named entity recognition in clinical notes. In *Proceedings of the workshop on biomedical information extraction*, pages 42–49. Association for Computational Linguistics.

Shaodian Zhang and Noémie Elhadad. 2013. Unsupervised biomedical named entity recognition: Experiments with clinical and biological texts. *Journal of biomedical informatics*, 46(6):1088–1098.

Guodong Zhou, Jie Zhang, Jian Su, Dan Shen, and Chewlim Tan. 2004. Recognizing names in biomedical texts: a machine learning approach. *Bioinformatics*, 20(7):1178–1190.

# Entity-Supported Summarization of Biomedical Abstracts

**Frederik Schulze and Mariana Neves**
Hasso-Plattner Institute
August-Bebel-Str. 88
Potsdam, 14482 Germany
mariana.neves@hpi.de

## Abstract

The increasing amount of biomedical information that is available for researchers and clinicians makes it harder to quickly find the right information. Automatic summarization of multiple texts can provide summaries specific to the user's information needs. In this paper we look into the use named-entity recognition for graph-based summarization. We extend the LexRank algorithm with information about named entities and present EntityRank, a multi-document graph-based summarization algorithm that is solely based on named entities. We evaluate our system on a datasets of 1009 human written summaries provided by BioASQ and on 1974 gene summaries, fetched from the Entrez Gene database. The results show that the addition of named-entity information increases the performance of graph-based summarizers and that the EntityRank significantly outperforms the other methods with regard to the ROUGE measures.

## 1 Introduction

There is an overload of textual information, also in the biomedical domain, where new research articles are published daily. Text summarization can support to deal with this textual data deluge by providing automatically generated summaries on certain topics, e.g., a gene or a disease, as well as supporting answers returned by question answering (QA) systems.

However, the adoption of these technologies in the biomedical domain is not straightforward. The domain specific language has different requirements for information extraction compared to news articles, where *who, when, what, and where* elements are often the most important. Additionally, there are less resources available in biomedicine for text summarization, such as benchmarking corpora or knowledge bases. Finally, requirements for summaries, such completeness or correctness, are even more important in this domain when compared to others, as important decision might be taken based on them. Therefore, text summarization for biomedicine raises new challenges that still need to be addressed.

Searching for specific information in biomedical publications is a hard task that involves screening many entries in PubMed [1], the most popular search engine in biomedicine. PubMed contains over 24 million records and is growing exponentially (Lu, 2011). Two thirds of all queries to PubMed return more than 20 results, which is probably the reason why 47% of all queries get followed by a subsequent query without accessing any abstract or article of the search result (Dogan et al., 2009). In average, users read four documents to find the information they search for. Text summarization can support this task by providing summaries of many publications for a certain query or topic. Automatic text summarization has also the potential to support database curation by automatically generating short summaries about a topic, such as the ones manually created for the Entrez Gene database.

In this work, we propose two graph-based summarization algorithms based on named entities for improving automatic multi-document summarization for the biomedical domain. While the first approach



[1] https://www.ncbi.nlm.nih.gov/pubmed



*Proceedings of the Fifth Workshop on Building and Evaluating Resources for Biomedical Text Mining (BioTxtM 2016)*,
pages 40–49, Osaka, Japan, December 12th 2016.

extends the lexical PageRank (Erkan and Radev, 2004) with information from the NER, the second approach is solely based on named-entity recognition (NER). Additionally, we show how to adapt these algorithms for particular uses cases in biomedicine by including a bonus score. We evaluate our methods in two uses cases: generation of ideal answers for question answering (QA) systems and for gene summaries.

The remainder of this paper is structured as follows: next section presents related work on text summarization. Section 3.2 introduces our summarization algorithms, followed by an evaluation of these in section 4. Finally, we present a discussion on the results, identify the limitations of our work, and propose future work in section 5.

## 2   Related work

Automatic text summarization has been studied since the late 1950s (Luhn, 1958) and has been applied to many domains such as news or social media, with some remarkable success (Chen et al., 2015). More recently, graph-based ranking algorithms like Google's PageRank  (Page et al., 1999) and the HITS algorithm  (Kleinberg et al., 1999) have been also successfully used for summarization. Essentially, these algorithms aim at deciding the importance of a vertex within a graph.

Mihalcea and Tarau introduced the TextRank (Mihalcea and Tarau, 2004) model, which essentially provides a general instruction for extracting lexical or semantical graphs from documents. In parallel and independent of the TextRank, Erkan and Radev introduced the LexicalPageRank (Erkan and Radev, 2004). One of the main changes that they proposed is the creation of multi-document summaries, by building a single sentence graph from multiple documents.

There are some previous work on summarization for biomedicine, such as (Kogilavani and Balasubramanie, 2009) that relied on ontologies to suport document retrieval and documents clustering. Researchers have also connected a graph-based extractive summarization algorithm with the domain knowledge of an ontology, for instance, for single document summarization (Morales et al., 2008). Their approach was similar to (Verma et al., 2007), but used a graph-based algorithm instead. Some researchers explored other non-traditional methods to generate automatic summaries. For example, generating not only summaries, but also a table of relevant data for extracting medical events and date times from documents (Aramaki et al., 2009). Another work proposed a system that uses patient data to provide a summarization of relevant information (Elhadad and McKeown, 2001).

Regarding summaries to support QA systems, some works focused on answering one special question type (e.g., "What is the best drug treatment for X?") (Demner-Fushman and Lin, 2006). Further, the general-purpose BioSquash system BIOSQUASH (Shi et al., 2007) was extended with biomedical domain knowledge from UMLS. The BioASQ challenge (Tsatsaronis et al., 2015; Krithara et al., 2016) was an important step to boost summarization solutions for biomedical QA systems, such as using machine learning approach based on Inductive Logic Programming (ILP) with different sets of features (Malakasiotis et al., 2015).

One of the first systems for automatic generation of gene summaries was proposed by (Ling et al., 2006). A different approach, that has some similarities to our work, was proposed by Jin et al. (Jin et al., 2009). They also use LexRank and extend it with two domain specific steps: identification of signature terms and calculation the similarity between each sentence based on the Gene Ontology (GO) terms. This approach is similar to our redundancy reduction step (cf. Section3.2). Finally, a similar approach to ours is the work of (Shang et al., 2014) which makes uses of TextRank based on frequency of words and LDA for topic relevance.

## 3   Methods and Materials

In this section, we introduce the data that we used and describe our methods for automatically generating summaries from scientific biomedical abstracts.

### 3.1 Data

Since we have two different use cases, we will also use two different datasets. However, both datasets rely on PubMed as source for the documents.

**PubMed.** PubMed [2] is a free search engine, that not only offers access to the MEDLINE database, but also to life science journals and online books. PubMed contains over 26 million records from over 26.000 journals.

**Domain dictionaries.** Our dictionaries combine data from the Unified Medical Language System[3] (UMLS), a collection of various health and biomedical vocabularies, and from SNOMED, a suite of standards from the U.S. Federal Government for the electronic exchange of health information. We specified a default name and a list of aliases and variations for each entity. Furthermore, entities were grouped by types, e.g., genes or diseases.

**EntrezGene.** EntrezGene[4] is a database for genes from various species. Each entry in EntrezGene is provided with a rich range of information[5], such as official symbol, corresponding organism and a short manually created summary.

### 3.2 Methods

In this section, we will describe the details of our summarization system and the algorithms that we have implemented.

**Document retrieval.** We fetched from PubMed the documents that we used for the generation of the summaries. We used the Entrez Programming Utilities [6], a set of public APIs that provide access to the data in PubMed. It allows users to get a full record by its PubMed identifier (PMID), as well as to retrieve documents which match some given keywords. Overall, we fetched 68.083 abstracts, which were used to generate summaries. We rely only on the abstracts, not on the full text, due to the following reasons: (a) most of the records in PubMed contain only an abstract; (b) sentences in an abstract are more suitable for a summary, since they are a summarization of an article and contains the most relevant information; (c) the BioASQ dataset (cf. section 4.1) are based on the abstracts, not on the full text.

**Document pre-processing.** We extracted linguistic and semantic annotations for the documents using the built-in text analysis functionality of an in-memory database (SAP HANA). Therefore, we created two full-text indexes (FTI), i.e., a a full indexing of all documents, one for linguistic annotations, i.e., part-of-speech (POS) tags and stems, and one for semantic annotations, i.e., named entities for genes, diseases, etc. The name-entity recognition (NER) was based on the custom dictionaries derived from various terminologies from UMLS, as previously described in our question answering system (Schulze et al., 2016). Dictionary matching to the documents was performed inside the database based on an approximated matching of the dictionaries with the documents. We mapped the words in the linguistic index to the entities in the NER index only for entities composed of a single word.

**Extended LexRank.** We extended the LexRank graph-based algorithm (Erkan and Radev, 2004) with information from the NER step. LexRank finds the most central sentences by building a sentence graph, based on the idf-modified cosine similarity, and calculates the PageRank on the resulting graph. We relied on the pre-processing step to normalize each sentence, based on the stemming and the named entities. After running the LexRank on these normalized sentences, we remove redundant sentences and extract the best sentences for a summary.

LexRank is based on the assumption that similar sentences contain the exact same words. However, two sentences can express the same content using different forms of the same word or even their syn-

---


[2] http://www.ncbi.nlm.nih.gov/pubmed
[3] http://www.nlm.nih.gov/research/umls/
[4] http://www.ncbi.nlm.nih.gov/gene
[5] e.g., humangeneHNF1A, http://www.ncbi.nlm.nih.gov/gene/6927
[6] http://www.ncbi.nlm.nih.gov/books/NBK25501/

onyms. We rely on stems, from the linguistic FTI, and on the named entities from the semantic FTI to normalize different but related words and to reduce the dimension of the vector space.

After merging both indices, we unify the resulting index by choosing a single form from one of the available ones. For instance, only 61% of the words were recognized as named entities. Therefore, we choose one form in this order: named entity, stem, normalized form or token text. Since the named entities not only include different forms of the same word but also synonyms and typical spelling mistakes, we chose this as the preferred form, followed by the stemmed form, that could also include different word classes and conjugations. The so-called normalized form only accounts for capitalization and mutated vowels, while the token text is the original word as it appears in the text. The unified index was used as input for the LexRank algorithm and for the calculation of the idf values.

After using LexRank to rank the sentences according to their centrality, we need additional post-processing steps to generate the summaries. Firstly, we extracted the sentences that are most suitable for a summary using the following process: (1) we initialized two sets: an empty set $A$ and a set $B$ that contains all extracted sentences; (2) we ordered the sentences in set $B$ by descending order according to their score; (3) we moved the top sentence $s_i$ from set $B$ to set $A$. Then penalized all sentences $s_j$ similar to $s_i$ by calculating their new score according to the equation 1 below, where $sim(s_i, s_j)$ is the similarity between two sentences, $t = 0.3$ is a threshold and $w = 0.5$ is a penalty factor; (4) we repeated steps 2 and 3 until enough sentences were in set $A$.

$$score(s_j) = \begin{cases} w \times score(s_j), & \text{if } sim(s_i, s_j) \geq t \\ score(s_j), & \text{otherwise} \end{cases} \tag{1}$$

At the end of this procedure, we obtained the most central sentences that are also as distinct as possible to each other. We concatenated these sentences to create a summary with the most relevant information.

**EntityRank.** We developed a second ranking approach inspired by the LexRank. Similar to LexRank, EntityRank also uses the similarity between the sentences to generate a sentence graph, but based on the named entities. Additionally, it also includes the possibility to adapt the calculation to the specific use cases.

Since the EntityRank is a graph-based algorithm, like LexRank and PageRank, we built a graph from the documents that we want to summarize. In order to represent the text as a graph, we created a similarity matrix by comparing every two sentences to each other, with no distinction between sentences that came from the same or from different documents. We experimented with two graph approaches, namely, weighted and non-weighted edges, and we implemented three approaches.

Our first approach used a non-weighted graph, like in LexRank, by adding non-weighted edges between the vertices that have a similarity greater than a certain threshold. The value of this threshold directly influences the density of the resulting graph, since a lower threshold results in more connections. After evaluating various values, we decided for 0.2. For the second approach, we created a weighted sentence graph. Compared to the unweighted graph, this method has no loss of information, since we add an edge between every two vertices whose corresponding sentences have a similarity greater than zero. This usually resulted in a much larger and dense graph, but it also contained much more information. We calculated the similarity between two sentences based on the cosine similarity and on the named entities. Our third hybrid approach combined the threshold used in the first method with the use of weighted edges from the second method. We add a weighted edge between every two sentences whose similarity score is larger than a certain threshold. We decided for a value of 0.1 for the threshold based on our experiments.

After creating the sentence graph, we calculated a score that represents the centrality of the sentence based on PageRank (Page et al., 1999), which is a round-based algorithm. We recalculated the score of every vertex in each round by using the results from the previous round until convergence of the scores, using the equation below:

$$\text{score}(s_i) = \frac{d}{N} + (1 - d) \sum_{s_j \in adj[s_i]} \frac{\text{score}(s_j)}{deg(s_j)} \tag{2}$$

where $N$ is the total number of vertices in the graph, $adj[s]$ are all adjacent vertices of the vertex $s$, $deg(s_j)$ is the degree of vertex $s_j$, and $d$ is a 'damping factor', which is typically set between 0.1 and 0.2 as proposed by (Page et al., 1999).

Since this formula is only defined on non-weighted graphs, we did not use it on weighted graphs. Instead, we modified it to distribute the score of each vertex, depending on the weights of its edges. The resulting equation is a modified version of the EntityRank for weighted graphs:

$$\text{score}(s_i) = \frac{d}{N} + (1 - d) \sum_{s_j \in adj[s_i]} \frac{cosine(s_i, s_j)}{\sum_{s_y \in adj[s_j]} cosine(s_y, s_j)} score(s_i) \tag{3}$$

where $cosine(s_1, s_2)$ is the cosine similarity between two sentences. With this formula, we calculated the non-weighted EntityRank and the weighted EntityRank using the same weight for every edge.

Similar to the PageRank, that gives more importance for certain Web sites, we introduced a score to describe the relevance of a sentence for a use case. This score was used to change the distribution of EntityRank and to give an higher weight to the sentences that have a greater relevance for the use case. We changed the equation accordingly:

$$\text{score}(s_i) = E(s_i)\frac{d}{N} + (1 - d) \sum_{s_j \in adj[s_i]} \frac{cosine(s_i, s_j)}{\sum_{s_y \in adj[s_j]} cosine(s_y, s_j)} score(s_i) \tag{4}$$

where $E(s_i)$ is the relevance of the sentence $s_i$ for the current use case. The average of this factor over all vertices is 1, in order to keep the average of all scores converging to 1.

**EntityRank for question answering.** Given that summaries for QA systems should focus on the question, we changed EntityRank to provide a bonus score for sentences that are related to the question. The bonus score was calculated based on the similarity of each sentence with the question, more specifically, on the common named entities. But before we can use the similarity, we need to normalize it, so that its average over all sentences is 1, by using the following equation:

$$E(s_i) = sim(s_i)\frac{|S|}{\sum_{j \in S} sim(s_j)} \tag{5}$$

where $S$ is the set of sentences and $sim(s_i)$ the similarity of the sentence $s_i$ to the question. We will use this normalized bonus score in the equation 4 to positively influence the summarization.

**EntityRank for gene summaries.** Since there is lot of information about most of the genes, the algorithm needs additional guidance on the right information for the summary. When analyzing human summaries from the Entrez Gene database, we noticed that they all cover similar topics, such as encoded protein, mutations, location or relation with certain diseases.

We created a bonus score that reflects how suited a sentence is for belonging to a gene summary. We relied on 9553 manually written summaries from EntrezGene to identify the most frequent named entities. The 15 most used terms were the following: "Proteins", "Genes", "protein location", "encoding", "variant", "family", "last name", "variant", "receptor", "receptor cells", "mutation", "numerous", "function", "enzymes", and "DNA". We created an artificial sentence from these most used terms, then we compared the artificial sentences to the sentences in the dataset. This similarity score was used to create the bonus score using equation 5 that was later applied in equation 4.

## 4 Evaluation

Unlike other domains, there are not gold-standard dataset for the evaluation of text summarization for biomedicine. Most papers introduce their own evaluation datasets, based either on abstracts from documents or on manual evaluation by experts. In this section, we evaluate our algorithms on the test collection provided by BioASQ and on a set of human summaries from the EntrezGene database. Tuning of parameters in our methods were solely based on the training sets.

### 4.1 Datatsets

**BioASQ**  We used 1009 questions and the corresponding data provided by the BioASQ challenge [7], an EU-funded project that aims to evaluate biomedical QA. These questions correspond to the datasets used in phase B of the tasks 1b (2013), 2b (2014) and 3b (2015). The BioASQ dataset includes not only the questions, but also the relevant documents (PMIDs from PubMed), passages and concepts, as well as exact answer and ideal answers (summaries). We relied on the these relevant documents (but not the passages) to generate our summaries. These datasets may include more than one ideal answer for each question, which we use for the evaluation of our system.

**EntrezGene summaries**  We automatically collected manual summaries from EntrezGene. The resulting summaries include some noise, therefore, we removed all summaries that were shorter than 100 characters and those which were equal for multiple genes. We then split this set randomly into a training set of 9553 summaries, that was used for generating the bonus score, and a test set of 1974 summaries. The quality of this dataset is very low, when compared to the BioASQ data. Further, the summaries have different lengths and some might be outdated. In the document retrieval step, we relied on the PubMed API to search for the official symbol, official full name and the whole name of the gene. We considered the top 20 abstracts for each gene as source for the summary.

### 4.2 Results

We compared our summarization algorithms to LexRank. Further, we also evaluated the performance of EntityRank for our two use cases. We used the Java implementation of ROUGE-N from the Dragon Toolkit (Zhou et al., 2003), which was developed by (Zhou et al., 2007). It implements ROUGE-1 and ROUGE-2 with additional stop word removal. Since ROUGE is a recall-based measure, the length has no influence on the score. Therefore, we calculated the ROUGE scores on summaries of similar length.

**Comparison to LexRank.**  We use the BioASQ dataset for comparison to LexRank, given its better quality. As recommended by BioASQ, we created summaries with a fixed length of 200 words and compared them to the reference summaries using both ROUGE-2 and ROUGE-1. Additionally, we also compared the algorithms for shorter summaries of only 100 words. Therefore, we extracted sentences until the next sentence would not fit in the limit of words.

Our comparison to LexRank is displayed in Figure 1. We compared LexRank to the extended LexRank, the non-weighted and the weighted EntityRank, without considering any bonus scores. In these diagrams, the blue bar shows the average score of all summaries, which adds up to 1009 summaries, while the green bar shows the average score only for summaries which were created from more than 20 documents, which adds up to 197 summaries. Results for ROUGE-2 (results not shown) had a similar correlation among the various systems, though lower results.

Our extended LexRank achieved a slightly better score, compared to the original LexRank. This was expected, since the algorithm is essentially the same. In contrast, the non-weighted and weighted variants of the EntityRank produced very different results. The overall score of the non-weighted EntityRank is far lower, compared to the weighted EntityRank. This can be explained by the loss of information that occurs when ignoring the similarity between the sentences. Finally, we got better ROUGE scores for the summaries that were created with the weighted EntityRank.

---

[7] www.bioasq.org

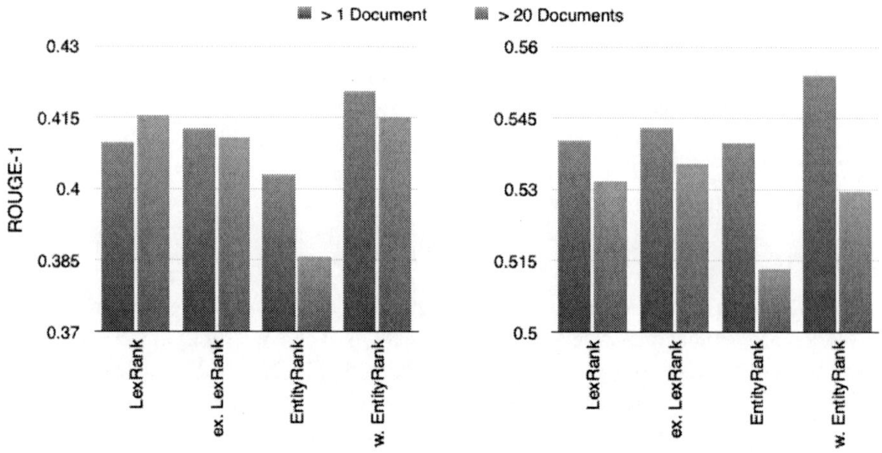


Figure 1: ROUGE-1 scores for 100-words summaries (left) and 200-words summaries (right).

**Question Answering.** We evaluated the influence of the bonus factor by comparing the adapted version of the weighted EntityRank with the general version. Therefore we generated 100- and 200-words summaries and compared them according to the number of abstracts that were used for the generation (cf. Figure 2). The bonus score slightly improved the results of the EntityRank, especially for summaries that were generated from fewer abstracts. While the 100-words summaries benefited from the bonus score, regarding summaries that were generated from less than 10 abstracts, 200-words summaries were improved or were similar for each number of abstracts. Although the overall improvement is only 2%, it shows that the bonus score did help guiding the EntityRank on the right sentences.

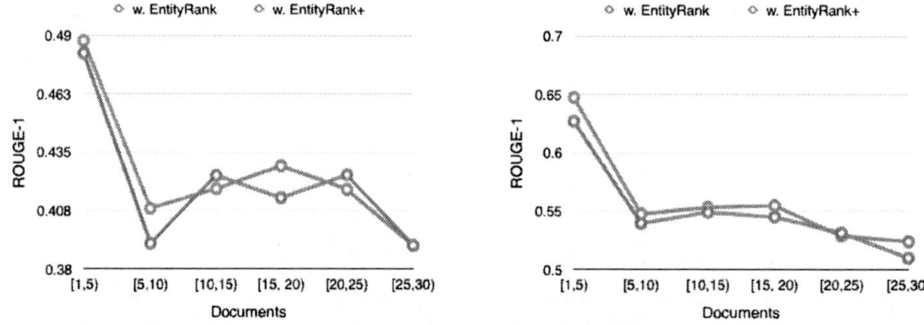


Figure 2: Comparison of ROUGE-1 scores for 100-words summaries (left) and 200-words summaries (right) of the weighted EntityRank (with and without the bonus score).

We also compared our systems to participants of the BioASQ challenge during the current fourth edition of the challenge (Schulze et al., 2016). We participated with the basic EntityRank and generated summaries with a length of five sentences instead of 200 words, which was the best version of our system that was ready at the time of the challenge (Spring/2016). As reported in (Schulze et al., 2016), we got a first position in one of the batches and good scores in the other batches.

**Gene Summarization.** We also evaluated whether our bonus score could improve the generation of gene summaries. We evaluated the same algorithms used earlier in the comparison to LexRank, as well as weighted and unweighted EntityRank with the bonus score (cf. Figure 3). The overall scores are significantly lower compared to the ROUGE-1 scores of the BioASQ dataset. This is due to the fact that the gene summaries is not a query-focused task and it can include many different topics related to the gene, while the summaries for QA should be related to the query (question). The weighted EntityRank obtained the highest scores among all systems. Thus, we can confirm the positive influence of the bonus score, which increased the scores of unweighted, as well as the weighted, EntityRank by roughly 2%.

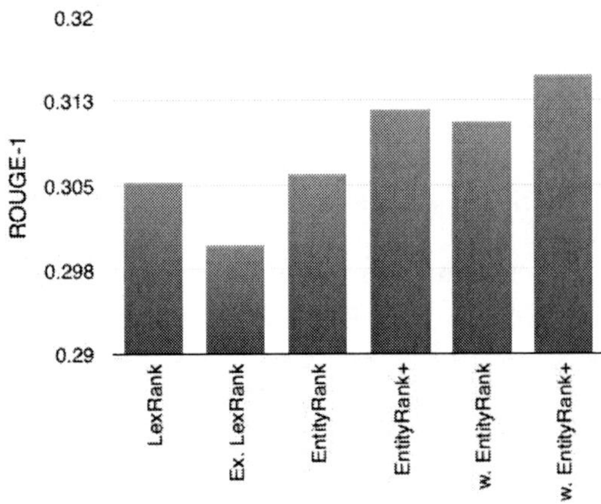


Figure 3: Comparison of ROUGE-1 scores for 200-words gene summaries.

## 5 Discussion and Future Work

Considering that summaries need dense sentences with much information, we chose to use only the abstracts of the PubMed articles. But especially for a focused summary, there could be information in the full paper that is not contained in the abstract. Therefore, in future work, we plan to evaluate the performance of our algorithms on full text, specially regarding the gene summaries, as information contained on these are not restricted to the abstracts.

The ordering of the sentences and the readability are important issue to create a fluent and natural summaries similar to human-written ones (Jurafsky and Martin, 2009). Further, it has a large impact on the comprehensiveness of a summary. For example, the information of a sentence could require knowledge of another sentence.

Further, we did not evaluate the performance of our NER approach for the detection of biomedical terms. The reason for not recognizing a term could not only be a failure of the NER step but could also be due to misspelling of words or missing terms in our dictionaries. Especially the latter could possibly occur more often during real use, if new documents contain words that is still not available in our dictionaries. Since our algorithms are heavily dependent on the named-entities, we need to check whether there is still room for improvement on the NER step, either regarding adding new terminologies or using machine learning approaches.

Not only false negatives are an issue, but also false positives. False positives can be either due to the NER algorithm or an error in the dictionaries. This is a problem specially for acronyms which often match common English words, such as conjunctions, when using an approximate and lowercase-based matching. A stopwords filtering step could remove some of these false positives.

Finally, one issue could raise from merging the linguistic and semantic indices. Set of words that have the same stem and a related meaning could have been handled as the same. But this was not the case if one of them was recognized as a named-entity. However, we anticipate that these were indeed rare cases.

## 6 Conclusions

Automatic text summarization has the potential to support many domains. It enable researchers to quickly get an overview of a specific topic, without investing too much time for searching the information. Especially for fast changing domains like biomedicine, it can help clinicians to save valuable time that could be use on treating patients. Therefore, summarization algorithms should utilize domain knowledge to create accurate summaries.

In order to improve the understanding of the domain, we relied on named-entity recognition in our summarization algorithms. We showed that adding named entities to graph-based summarization algorithm did improve the results for the task. The named-entity information supported building a sentence graph, by improving the similarity measure. This approach is especially effective for summaries that are generated from few documents.

We showed two different ways of enriching graph-based algorithms with named-entity information. The extended LexRank proved that using named entities greatly improves the overall results of the LexRank in the medical domain. But the main contribution of this work is EntityRank, which is a graph-based algorithm that is solely based on named entities instead of terms. We showed that a graph-based summarization algorithm that only uses technical terms can outperform other graph-based approaches within a restricted domain. Although the performance only increases for summaries that are created from few documents, it still shows the potential of that approach, since it performs comparable to other state-of-the-art systems.

Additionally, we showed how the EntityRank could be adapted to more specialized use cases, such as question answering and gene summarization. We implemented two bonus scores and showed how they improve the results for special use cases.


## References


Eiji Aramaki, Yasuhide Miura, Masatsugu Tonoike, Tomoko Ohkuma, Hiroshi Mashuichi, and Kazuhiko Ohe. 2009. Text2table: Medical text summarization system based on named entity recognition and modality identification. In *Proceedings of the Workshop on Current Trends in Biomedical Natural Language Processing*, pages 185–192. Association for Computational Linguistics.

Kuan-Yu Chen, Shih-Hung Liu, Berlin Chen, Hsin-Min Wang, Ea-Ee Jan, Wen-Lian Hsu, and Hsin-Hsi Chen. 2015. Extractive broadcast news summarization leveraging recurrent neural network language modeling techniques. *Audio, Speech, and Language Processing, IEEE/ACM Transactions on*, 23(8):1322–1334.

Dina Demner-Fushman and Jimmy Lin. 2006. Answer extraction, semantic clustering, and extractive summarization for clinical question answering. In *Proceedings of the 21st International Conference on Computational Linguistics and the 44th annual meeting of the Association for Computational Linguistics*, pages 841–848. Association for Computational Linguistics.

Rezarta Islamaj Dogan, G Craig Murray, Aurélie Névéol, and Zhiyong Lu. 2009. Understanding pubmed® user search behavior through log analysis. *Database*, 2009:bap018.

Noemie Elhadad and Kathleen R McKeown. 2001. Towards generating patient specific summaries of medical articles. In *In Proceedings of NAACL-2001 Workshop Automatic*. Citeseer.

Günes Erkan and Dragomir R Radev. 2004. Lexrank: Graph-based lexical centrality as salience in text summarization. *Journal of Artificial Intelligence Research*, pages 457–479.

Feng Jin, Minlie Huang, Zhiyong Lu, and Xiaoyan Zhu. 2009. Towards automatic generation of gene summary. In *Proceedings of the Workshop on Current Trends in Biomedical Natural Language Processing*, pages 97–105. Association for Computational Linguistics.

Dan Jurafsky and James H. Martin. 2009. *Speech and Language Processing: An Introduction to Natural Language Processing, Computational Linguistics, and Speech Recognition*. Pearson international edition. Prentice Hall.

Jon M Kleinberg, Ravi Kumar, Prabhakar Raghavan, Sridhar Rajagopalan, and Andrew S Tomkins. 1999. The web as a graph: measurements, models, and methods. In *Computing and combinatorics*, pages 1–17. Springer.

AA Kogilavani and B Dr P Balasubramanie. 2009. Ontology enhanced clustering based summarization of medical documents. *International Journal of Recent Trends in Engineering*, 1(1).

Anastasia Krithara, Anastasios Nentidis, Georgios Paliouras, and Ioannis Kakadiaris. 2016. Results of the 4th edition of bioasq challenge. In *Proceedings of the Fourth BioASQ workshop at the Conference of the Association for Computational Linguistics*, pages 1–7.

Xu Ling, Jing Jiang, Xin He, Qiaozhu Mei, Chengxiang Zhai, and Bruce Schatz. 2006. Automatically generating gene summaries from biomedical literature.

Zhiyong Lu. 2011. Pubmed and beyond: a survey of web tools for searching biomedical literature. *Database*, 2011:baq036.

Hans Peter Luhn. 1958. The automatic creation of literature abstracts. *IBM Journal of research and development*, 2(2):159–165.

Prodromos Malakasiotis, Emmanouil Archontakis, Ion Androutsopoulos, Dimitrios Galanis, and Harris Papageorgiou. 2015. Biomedical question-focused multi-document summarization: Ilsp and aueb at bioasq3. In *Working Notes for the Conference and Labs of the Evaluation Forum (CLEF), Toulouse, France*.

Rada Mihalcea and Paul Tarau. 2004. Textrank: Bringing order into texts. Association for Computational Linguistics.

Laura Plaza Morales, Alberto Díaz Esteban, and Pablo Gervás. 2008. Concept-graph based biomedical automatic summarization using ontologies. In *Proceedings of the 3rd textgraphs workshop on graph-based algorithms for natural language processing*, pages 53–56. Association for Computational Linguistics.

Lawrence Page, Sergey Brin, Rajeev Motwani, and Terry Winograd. 1999. The pagerank citation ranking: bringing order to the web.

Frederik Schulze, Ricarda Schler, Tim Draeger, Daniel Dummer, Alexander Ernst, Pedro Flemming, Cindy Perscheid, and Mariana Neves. 2016. Hpi question answering system in bioasq 2016. In *Proceedings of the Fourth BioASQ workshop at the Conference of the Association for Computational Linguistics*, pages 38–44.

Yue Shang, Huihui Hao, Jiajin Wu, and Hongfei Lin. 2014. Learning to rank-based gene summary extraction. *BMC bioinformatics*, 15(Suppl 12):S10.

Zhongmin Shi, Gabor Melli, Yang Wang, Yudong Liu, Baohua Gu, Mehdi M Kashani, Anoop Sarkar, and Fred Popowich. 2007. Question answering summarization of multiple biomedical documents. In *Advances in Artificial Intelligence*, pages 284–295. Springer.

George Tsatsaronis, Georgios Balikas, Prodromos Malakasiotis, Ioannis Partalas, Matthias Zschunke, Michael R Alvers, Dirk Weissenborn, Anastasia Krithara, Sergios Petridis, Dimitris Polychronopoulos, et al. 2015. An overview of the bioasq large-scale biomedical semantic indexing and question answering competition. *BMC bioinformatics*, 16(1):138.

Rakesh Verma, Ping Chen, and Wei Lu. 2007. A semantic free-text summarization system using ontology knowledge. In *Proc. of Document Understanding Conference*.

Xiaohua Zhou, Xiaodan Zhang, and Xiaohua Hu. 2003. The Dragon Toolkit. `dragon.ischool.drexel.edu` [Online; accessed 14-June-2016].

Xiaohua Zhou, Xiaodan Zhang, and Xiaohua Hu. 2007. Dragon toolkit: incorporating auto-learned semantic knowledge into large-scale text retrieval and mining. In *Tools with Artificial Intelligence, 2007. ICTAI 2007. 19th IEEE International Conference on*, volume 2, pages 197–201. IEEE.

# Fully unsupervised low-dimensional representation of adverse drug reaction events through distributional semantics

**Alicia Pérez, Arantza Casillas, Koldo Gojenola**
IXA research group (`http://ixa.si.ehu.eus`)
University of the Basque Country (UPV-EHU)
`{alicia.perez,arantza.casillas,koldo.gojenola}@ehu.eus`

## Abstract

Electronic health records show great variability since the same concept is often expressed with different terms, either scientific latin forms, common or lay variants and even vernacular naming. Deep learning enables distributional representation of terms in a vector-space, and therefore, related terms tend to be close in the vector space. Accordingly, embedding words through these vectors opens the way towards accounting for semantic relatedness through classical algebraic operations.

In this work we propose a simple though efficient unsupervised characterization of Adverse Drug Reactions (ADRs). This approach exploits the embedding representation of the terms involved in candidate ADR events, that is, drug-disease entity pairs. In brief, the ADRs are represented as vectors that link the drug with the disease in their context through a recursive additive model.

We discovered that a low-dimensional representation that makes use of the modulus and argument of the embedded representation of the ADR event shows correlation with the manually annotated class. Thus, it can be derived that this characterization results in to be beneficial for further classification tasks as predictive features.

## 1 Introduction

The aim of this work is to represent Adverse Drug Reactions (ADRs) efficiently so as to find drug related etiologies in Electronic Health Records (EHRs). Nebeker et al. (2004) defined an ADR as "a response to a drug which is noxious and which occurs as doses normally used". Finding ADRs efficiently is of much concern to pharmaco-surveillance and clinical documentation services. Personnel at pharmaco-surveillance services reads thousands of EHRs in order to detect this type of events and, furthermore, documentation services claim that, while ADRs should be reported by law, they seem to be under-reported (Dalianis et al., 2015).

From the natural language processing perspective, EHRs differ substantially from clinical literature as PubMed (Cohen and Demner-Fushman, 2014) in aspects such as syntax, the use of non-standard abbreviations (Okazaki et al., 2010; Kreuzthaler and Schulz, 2015), and misspellings (Dalianis, 2014). Within EHRs it is common to find the same concept expressed with different terms or surface-forms, synonyms or near-synonyms, either scientific latinised forms, common or lay variants or even vernacular naming, misspells, abbreviations etc. For example, we have found in the Spanish corpus we are working with a wide variety of ways to refer to the diagnostic term with code 600.00 from the ninth Clinical Modification of the International Classification of Diseases (WHO, 2014), namely, "benign prostatic hyperplasia" e.g. *hipertrofia benigna de próstata, hiperplasia BP, HBP-II, hiperplasia benigna de la prstata en estado II*, etc.

Distributional semantics has demonstrated to be a powerful approach to represent closely in a continuous space ($\mathbb{R}^n$) related entities. The rationale is to represent similar entities by means of close points in that space (or word-embedding) since close points render related meaning. Hence, embedding words





*Proceedings of the Fifth Workshop on Building and Evaluating Resources for Biomedical Text Mining (BioTxtM 2016),*
pages 50–59, Osaka, Japan, December 12th 2016.

through vectors opens the way towards semantic search (Thompson et al., 2016) as an alternative to the classical string-based methods that rely on keyword search. In this context, distributional semantics arises as a naturally appropriate framework as it exploits semantic similarity rather than the frequency of target surface-forms (Batista-Navarro et al., 2016). Deep learning (LeCun et al., 2015) has proven useful to model semantic relatedness based on corpus (Mikolov et al., 2013b). It accounts contextual information and patterns that occur in big amounts of unsupervised corpora to create sketches of word-forms that embed semantic relatedness. Mikolov et al. (2013a) proposed a distributed word embedding model that allowed to convey meaningful information on vectors derived from neural networks. With this approach, semantically related terms tend to be close in the vector space and the advantage is that closeness is a well studied metric in vector spaces (e.g. through Euclidean distance) while measuring semantic closeness is not trivial.

Our work is inspired by the example proposed by Mikolov et al. (2013a), where the authors present the association of the following terms: *Madrid* is to *Spain* what a *query* entity is to *France* or more formally as in (1).

$$\overrightarrow{Spain - Madrid} \approx \overrightarrow{France - query} \tag{1}$$

Here, each entity as *Madrid*, *Spain* or *France*, is represented as a point in a vector space that conveys meaningful information. The intuition is that close points correspond to semantically related entities. Accordingly, the hypothesis stands that similar vectors constructed by linking points also convey similar relations. Bearing this in mind and back to our domain, our research question is as follows: are drug-related aetiologies similarly represented in a semantic space? That is, given that a disease (e.g. "nosebleed") was caused by a drug (e.g. "sintrom") in a given EHR, can we extract other relations for their location in the vector space? Moreover, are similar the vectors that trigger ADR events and can be distinguished from those that do not form ADR events? We have tried to state this question formally through expression (2), where we denoted an ADR candidate by a disease-drug pair and, particularly, denoted as $\oplus$ the ADR events, $\ominus$ the non-ADR events and the sub-indices ($_i$ and $_j$) simply refer to a given particular instance in the data. We have tried to depict the research question through Figure 1 which shows relevant entities as points in a space and also a few ADR events through vectors linking drugs and diseases.

$$\overrightarrow{(Disease - Drug)}_i^{\oplus} \overset{?}{\approx} \overrightarrow{(Disease - Drug)}_j^{\oplus} \tag{2}$$

$$\overrightarrow{(Disease - Drug)}_i^{\oplus} \overset{?}{\napprox} \overrightarrow{(Disease - Drug)}_j^{\ominus}$$

The contribution of this paper is an efficient representation of ADR events with high correlation with the class ($\mathcal{C} = \{\ominus, \oplus\}$). The interest behind stands in its potential use for further supervised classification tasks as a stand-alone technique or, as it is our purpose, as predictive features for other classification techniques. All in all, we focus on representation while classification is out of the scope of this paper.

## 1.1 Related work

A big challenge of ADR event extraction is the fact that ADRs represent rare or infrequent events. In real EHRs we saw that ADRs represent 1% of the drug-disease pairs. That is, the ADR event extraction task is significantly skewed towards the negative class ($\ominus$ non-ADR events) in real EHRs and, hence, it represents a complex and still open problem for supervised classification. Regarding this issue, Henriksson (2015) created an artificially balanced corpus consisting of positive examples, health-care episodes coded with ADR-related diagnosis codes, and the negative examples were an equal number of randomly selected examples. By contrast, in this work we tackle the ADR extraction problem in its natural context, without avoiding the data skewness problem. Accordingly, we explore the scope of the proposed representation in the real context.

Last years authors have combined both supervised and unsupervised techniques to tackle entity recognition (Agerri and Rigau, 2016) and event extraction (Zhang et al., 2015; Zhou et al., 2015). Focusing on the clinical domain, the task presented by Henriksson et al. (2015) entailed the detection of health-care episodes that involved an ADR. They were pioneers in representing health-care episodes using semantic

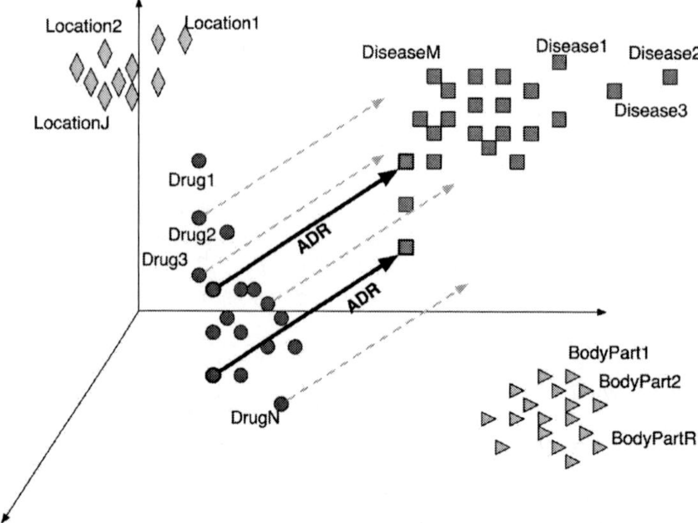


Figure 1: We represent the entities together with their context by means of word-embeddings (e.g. blue dots represent drugs, red squares diseases, triangles body-parts etc) and the ADR candidates through vectors linking a drug with a disease. The goal is to explore if the disposition of the ADR events is similar in the distributional semantic space.

spaces, that is, they extracted different representations of the documents and a semantic space was created for each component of the representation. The semantic spaces and other features were used as input to a machine learning algorithm. The task presented in (Henriksson et al., 2015) differs from ours since ADRs were a sub-set of diagnostic terms referred to as ADRs in the ICD-10-CM. This subset consisted of particular diseases that always had a drug as their cause. The goal was not to relate two entities (a drug and a disease) but, instead, it was to recognize a sub-set of disease-entities (a sub-set of ICDs). By contrast, we aim at extracting relations between drug and disease entities rather than subsets of entities.

Typically event extraction tasks and challenges (Pradhan et al., 2014) focus on the extraction of events that occur within the same sentence. Trying to relate entities that are in different sentences is by far much more complicated due to the amount of information that is required to take into account as the distance increases, not to mention anaphora and co-reference resolution. Nevertheless, the systems that only attempt at finding intra-sentence events might be discarding valuable information as many relations involve entities in different sentences. Indeed, in our set of EHRs, the inter-sentence events represent 51.7% of the positive instances, besides, on average the positive events are placed at a distance of 4 sentences but we found positive events involving entities further than 15 sentences. With the representation of each drug-disease pair proposed in this work we do not restrict ourselves to explore only intra-sentence events (as it tends to be the main trend in this area) but we also cope with inter-sentence events.

## 2   Methodology: ADRs as relation-vectors

Full classification systems rely entirely on predictive features to infer a model. The features represent, hence, a crucial source of knowledge. Our aim is to get an efficient representation of ADR events. By virtue of distributional semantics we prove that relevant features can be obtained. In an attempt to build a model able to extract events, the event should be characterized in an efficient manner. The key issue is that the representation itself should show correlation with the type of event (e.g. $\oplus$ for ADR events and $\ominus$ for non-ADR events).

We resort to semantic vector spaces to represent the entities (drugs and diseases), that is, each entity will be represented by its vector, calculated by the word2vec tool (Mikolov, 2016). As a result, each word has associated a point in an $\mathbb{R}^n$ vector-space. Nevertheless, a given drug does not always provoke the same side effects, hence, the context turns out crucial to set the relations. Accordingly, we propose to encompass the entity together with its right and left contexts (being the contexts within a window of size $m$). Note that even though word vectors trained by means of Skip-gram model encode context information in the corpus, this context information reveals the global trend of the sample. With the general contextual trend, the model assigns a location in the space to each word. Nevertheless, the rationale behind the use of the context is that focusing on each EHR can leverage the local contextual information. In this regard, given an EHR in its textual form: $x^{(1)}, x^{(2)}, \cdots, x^{(i)}, \cdots, x^{(s)}$, and given that we identified $x^{(i)}$ as an entity (either drug or disease) that shall be taken into account as a candidate that could trigger an ADR event, the issue is how to render all the contextual information information as well. To do so, it is common practice to turn to the embedding of each word, that is, for a given word in the vocabulary, $x^{(k)} \in \Sigma$, get the corresponding embedding $\vec{x}^{(k)} = (x_1^{(k)}, x_2^{(k)}, \cdots, x_n^{(k)}) \in \mathbb{R}^n$ where $1 \leq k \leq |\Sigma|$ and $n$ is the pre-fixed size for the dimension of the space. Nevertheless, in this work we do not rely only on their associated embedding but on a vector formed as a linear combination of the context vectors as stated through (3). In brief, for a given word $x^{(i)} \in \Sigma$, we get a vector $\vec{x}^{(i)} \in \mathbb{R}^n$, but in such a way that it comprises not only the word but its context as well, that is, a window of $m$ words to the left and to the right.

$$v\big(x^{(i-m)}, x^{(i-m+1)}, \cdots, x^{(i)}, \cdots, x^{(i+m-1)}, x^{(i+m)}\big) \quad = \quad \sum_{k=-m}^{m} \lambda_k \vec{x}^{(i+k)} \tag{3}$$

Expression (3) represents a *recursive additive model* (Ferrone and Zanzotto, 2013) that makes a representation for each word within a given context, but adapted to the entity on the focus through a context of a given length ($m$). The weight $\lambda_k$ balances the contribution of the context-word $k$ in the representation, as an alternative to the *basic additive model* (Mitchell and Lapata, 2008; Zanzotto et al., 2010).

For instance, given that we would like to explore whether the disease $x^{(i)}$ was caused by the drug $y^{(j)}$ in a given EHR, first we represent the entities following the recursive additive representation. That is, for the word $x^{(i)}$ we get the vector $v(x^{(i-m)}, x^{(i-m+1)}, \cdots, x^{(i)}, \cdots, x^{(i+m-1)}, x^{(i+m)})$ that, for the sake of brevity, shall be referred as $\mathbf{x}$; likewise, for $y^{(j)}$ we get its contextual recursive representation $\mathbf{y}$. Given the representation in an $n$-dimensional semantic space of a disease, $\mathbf{x} \in \mathbb{R}^n$, and of a drug, $\mathbf{y} \in \mathbb{R}^n$, our goal is to carry out data analysis and measure if this characterization helps to represent the relatedness of those concepts and, thus, assess quantitatively if the semantic space helps to guess if the pair is an ADR event or not.

Quite naturally, we defined the contextual *relation vector* as the vector that starts in the point $\mathbf{x}$ and ends in $\mathbf{y}$ to represent the ADR. Note that $\overrightarrow{\mathbf{xy}} = \mathbf{y} - \mathbf{x}$, thus, $\overrightarrow{\mathbf{xy}} \in \mathbb{R}^n$. We explored both the *cosine similarity* between the entities $\mathbf{x}$ and $\mathbf{y}$ and also the *euclidean distance* between them or, what is equivalent, the argument and modulus of the relation vector $\overrightarrow{\mathbf{xy}}$. Cosine similarity is formally stated in (4) and the euclidean distance in (5), both of them are graphically depicted in Figure 2. It is well worth mentioning that $x_i$ in (5) represents the $i$-th component of vector $\mathbf{x}$ and likewise, $y_i$ for $\mathbf{y}$.

$$\cos(\theta) \quad = \quad \frac{\mathbf{x} \cdot \mathbf{y}}{|\mathbf{x}||\mathbf{y}|} \tag{4}$$

$$d(\mathbf{x}, \mathbf{y}) \quad = \quad |\overrightarrow{\mathbf{xy}}| = \left[ \sum_{i=1}^{n} (x_i - y_i)^2 \right]^{\frac{1}{2}} \tag{5}$$

Note that the smaller the argument $\theta \equiv \angle(\mathbf{x}, \mathbf{y})$, the bigger its cosine and, thus, the more related the entities associated to $\mathbf{x}$ and $\mathbf{y}$. In the same way, the smaller the modulus $|\overrightarrow{\mathbf{xy}}|$, the more related the drug and disease entities. This is the reason for which we opted to provide the results in terms of the argument and the euclidean distance, both decrease as the entities get related.

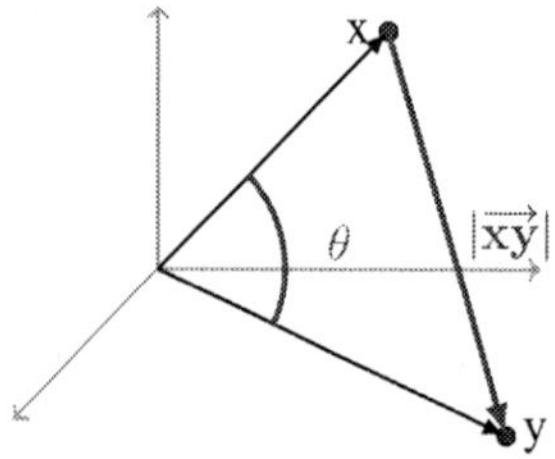

Figure 2: Argument $\theta$, between vectors $\mathbf{x}$ and $\mathbf{y}$, and modulus of the relation vector $\vec{\mathbf{xy}}$

Our aim is to analyze if the event-vectors $\vec{\mathbf{xy}}$ that relate ADRs show similarities (and likewise for non-ADRs). In other words, we aim at analyzing two sets of event-vectors that contain, respectively, ADR (or positive events) and non-ADR (or negative events), denoted as $\mathcal{S}_\oplus$ in (6) where $c(\vec{\mathbf{xy}})$ represents the real class of the event (equivalently, for $\mathcal{S}_\ominus$).

$$\mathcal{S}_\oplus = \{\vec{\mathbf{xy}} : c(\vec{\mathbf{xy}}) = \oplus\} \tag{6}$$

There are more sophisticated approaches that represent relation extraction through directed graphs that also make use of continuous vector spaces (Bordes et al., 2013). This kind of multi-relational data can represent entities as points in a vector space and relations as operations, such as projections, translations, etc. (Wang et al., 2014). In our work we explore a simple approach in a particular context where the relations are highly unbalanced. Nevertheless, relational machine learning approaches (Nickel et al., 2016) have demonstrated an for their ability to model and generalize relations. In particular, constraining the embedding in such a way that semantically related entities were placed in a lower-dimensional subspace (Jameel and Schockaert, 2016) seems applicable to our task. Provided that drug and disease entities would be represented in different subspaces while each sub-space would ensure that drug families would still be distinguished.

## 3 Experimental results

### 3.1 Task and corpus

We focus on EHRs written in Spanish by staff from the Galdakao-Usansolo and Basurto Hospitals. Admittedly, getting this kind of corpora to do research is not easy due to confidentiality issues (Cohen and Demner-Fushman, 2014), and it is even more difficult when it comes to explore other languages rather than English (Névéol et al., 2014). All in all, as Spanish is official in many countries, developing clinical text mining results for this language is of much interest, not only for the health systems but also for patients so that they get their EHRs in their own language.

The analysis of the proposed representation for relation vectors was built up based on an unsupervised or unannotated corpus, an in-domain medium-sized unsupervised set formed by 141,000 EHRs. From this partition we computed the word embeddings ($\vec{\mathbf{x}}^{(k)}$) that served to get the two features that are being proposed in this article (namely, $\theta$ and $|\vec{\mathbf{xy}}|$). Next, the assessment of the proposed representation was carried out on two independent supervised test sets not contained within the unsupervised set. In other words, we aimed to measure the correlation between $\theta$ and $|\vec{\mathbf{xy}}|$ with respect to the class ($\mathcal{C} = \{\oplus, \ominus\}$). The total number of tokens and documents involved in each set are shown in Table 1. To sum up, the

| | tokens | docs | $|\mathcal{S}_\oplus|$ | $|\mathcal{S}_\ominus|$ |
|---|---|---|---|---|
| unsupervised | $52 \times 10^6$ | 141,000 | - | - |
| test-1 | $21 \times 10^3$ | 41 | 58 | 21,911 |
| test-2 | $11 \times 10^3$ | 17 | 38 | 17,654 |

Table 1: Data sets: unsupervised to train the word embeddings and two supervised sets for testing.

resources exploited for the detection of ADRs are simply based on an unsupervised corpus, since we turn to word-embeddings and, from them, we derive the two proposed features ($\theta$ and $|\overrightarrow{\mathbf{xy}}|$). For the sake of curiosity, in this task we got an inter-annotator agreement of 82.86% on the test sets.

An inspection to Table 1 reveals a challenging (Monard and Batista, 2002; Phua et al., 2004; Mu et al., 2010) characteristic intrinsic of ADR detection: the classes are highly skewed being $\ominus$ the majority class. There are many works in this field that tackle ADR extraction with artificially balanced test sets (Henriksson, 2015). By contrast, we keep the repetition ratio as it is in the original sample of EHRs. Our aim is to check if this technique would help in real practice.

## 3.2  Results

The data analysis based on the proposed relation vector for ADR and non-ADR events is shown in Table 2. It presents the average argument $\theta$ and euclidean distance of relation vectors in each set ($\mathcal{S}_{\oplus}$ and $\mathcal{S}_{\ominus}$). Regarding the configuration of word2vec, we employed the skip-gram choice and a window of size s=5 requiring a n=300 dimensional vector space trained on the unsupervised set. Next, in order to represent each entity through the recursive additive model proposed in expression (3), a symmetric context of $m = 3$ tokens was chosen. Besides, we considered all the elements within the window equally influent and, hence, decided for $\lambda_i = 1$. Needless to say, this experimental setup involved a series of parameters ($s$, $n$, $m$, $\lambda$) that could have been fine-tuned on the basis of a supervised training corpus. Such a tuning would have helped to reassure the influence of each of them, for instance, with $m$ an empirical comparison of the need to exploit local context and its scope when it comes to get the relation vector; with $\lambda_i$ the influence of the context as the scope increases. While these empirical comparisons are of interest we found them out of the scope of this work.

|  | $\theta$ | | $|\overrightarrow{\mathbf{xy}}|$ | |
|---|---|---|---|---|
|  | mean | stdev | mean | stdev |
| $\mathcal{S}_{\oplus}$ | 1.10 | 0.26 | 10.89 | 2.98 |
| $\mathcal{S}_{\ominus}$ | 1.35 | 0.09 | 15.05 | 2.70 |

Table 2:  Argument and modulus of ADR candidate events.

Figure 3 shows that ADR and non-ADR events from test-1 represented as relation vectors are statistically different in terms of both $\theta$ and $|\overrightarrow{\mathbf{xy}}|$. Hence, any of the proposed features ($\theta$ and $|\overrightarrow{\mathbf{xy}}|$) allows to distinguish between ADR and non-ADR events.

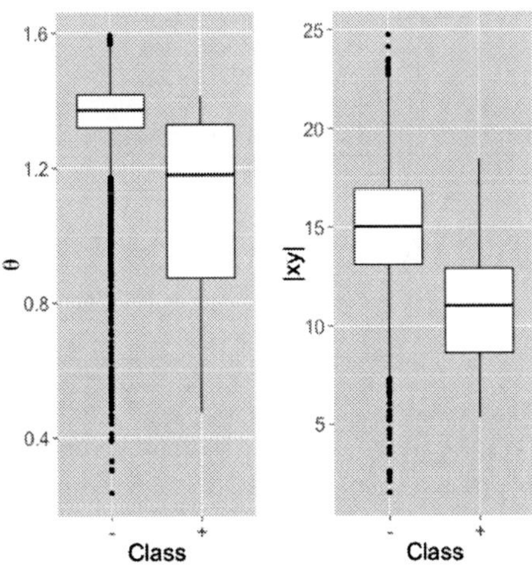


Figure 3:  Box-plot for the argument, $\theta$, and modulus, $|\overrightarrow{\mathbf{xy}}|$ with respect to the class of the event

The reader might wonder what would happen with an independent test from the same domain. We tackled this question through test-2 (presented in Table 1) and observed that the representation remains as stable as for the test-1. Indeed, both $\theta$ and $|\overrightarrow{\mathbf{xy}}|$ remained as in Table 2. Note that the model follows the intuition that related ADRs pose both small $\theta$ and small $|\overrightarrow{\mathbf{xy}}|$. In addition, one-way analysis of variance in the modulus stated that for this data-set, with respect to the null hypothesis of equal modulus, the p-value is $3.7 \times 10^{-15}$. The same applies to $\theta$ with a p-value of the same order of magnitude.

The proposed model is able to deal with inter-sentence and intra-sentence events. We wondered if the proposed event-vector representation gets degraded as the distance in sentences between the drug-disease pairs increases. In other words, does the event-vector remain stable in the vector-space despite they are in different sentences? Figure 4 comes to answer these questions. We explored the modulus of the relation vector for negative and positive instances. To be precise, we turned to the *relative location* of the drug $x^{(i)}$ with respect to the disease $y^{(j)}$ measured in sentences: $location(x^{(i)}, y^{(j)}) \equiv numSent(y^{(j)}) - numSent(x^{(i)})$. Whenever the drug and the disease are in the same sentence this location is 0; else, if the disease precedes the drug in the document, then the relative location is positive; otherwise, it is negative. Figure 4 depicts the box-plot associated to $|\overrightarrow{\mathbf{xy}}|$ for each class on three different location-ranges. We noticed that the regular way of reporting ADRs in EHRs in Spanish follows a scheme where the relative location is negative, and for them, $|\overrightarrow{\mathbf{xy}}|$ turned out very helpful to discriminate the class. As the trend changes, and particularly for very long positive relative locations, see the range "$(5, Inf]$" in Figure 4, the correlation of $|\overrightarrow{\mathbf{xy}}|$ with respect to the class decreases. We conclude that $|\overrightarrow{\mathbf{xy}}|$ is a helpful discriminant feature for ADR classification, particularly for those events that occur within the same sentence or relatively close (within 5 sentences), but also for the events that follow the trend and show a negative relative location despite of being far from each other.

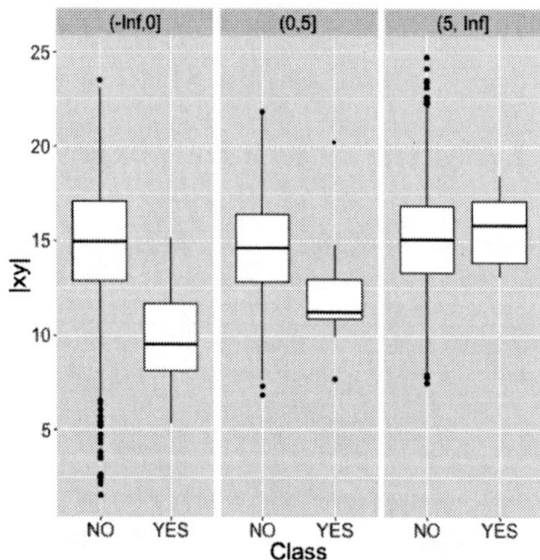


Figure 4: Impact on $|\overrightarrow{\mathbf{xy}}|$ of the relative location, in sentences, from the disease to the drug.

Even more, we carried out preliminary experiments with simple classifiers and, while it is out of the scope of this paper, the results are consistent with the hypotheses. The proposed features helped to discriminate ADR events. An example of the ADRs detected in a real EHR are shown in Figure 5 through Brat (Stenetorp et al., 2012). In this figure, the drug entities are marked in green with the tag `Grp_Medicamento` while the disease entities are marked in green with the tag `Grp_Enfermedad`. Positive ADR events are linked through arrows. In these examples all the ADRs occur within the same sentence, note that intra-sentence events.

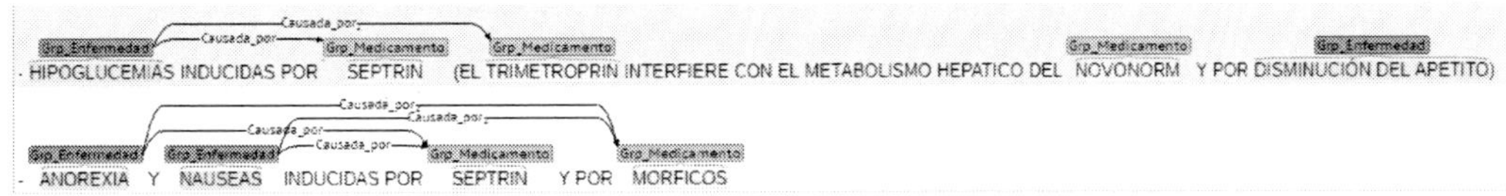



Figure 5: Example of ADRs detected in a real EHR.

In the mid-term we mean to test these features with a wide-suite of classifiers and focus, entirely, on classification techniques and results rather than on representation which is, indeed, the focus of this paper. Hence, we conclude that empirical observations lead us to support the representation introduced.

## 4 Concluding remarks and future work

In this work we proposed an efficient vector representation of drug-disease pairs, or ADR events, derived from distributional semantics. This representation of events showed clear correlation with respect to the class ($\oplus$ for ADR events and $\ominus$ for non-ADR events). In plain words, we made a linear combination of the context-vectors of each entity (either drug or disease) to get the vector representation of the entity within its context. Next, we formed the event by linking both entities yielding a relation vector.

Accordingly, we propose the use of a contextual recursive additive model to characterize each entity, either drug (represented as $x$) or disease $y$. Other approaches could have been explored, such as the mean vector (instead of the sum) of the context-vectors. Next, we related the drug trough the relation vector defined as $\overrightarrow{xy}$. From here, we proved that two characteristics derived from this relation vector ($\theta$ and $\overrightarrow{xy}$) showed clear correlation with respect to the class of the event ($\oplus$, $\ominus$). This work does not aim at proposing this low-dimensional representation as a stand-alone ADR event extraction technique, by contrast, we think of this as a prior step in order to leverage the representation of ADRs for subsequent supervised classification methods. This representation settled a basis for an ongoing work focused on the ADR classification.

Even though distributional semantics is known for its ability to embed related words in close positions of the vector space, there are still open challenges. Limitations of the approach explored in this paper stand on that we do not cope with ADRs expressed by means of aphoristic pronouns and co-referent expressions (such as "it"), even though distributional semantics could approach those terms for their co-appearance as well.

Future work is planned in two directions: on the one hand we aim at going ahead and try fully unsupervised classification of ADR events by improving this representation and enhancing it with LDA analysis; in parallel, we shall focus on feeding supervised classifiers with this approach and experimenting thoroughly if this low-dimensional vector representation can leverage the performance of current supervised methods.

## Acknowledgments


The authors would like to thank the personnel of Pharmacy and Pharmacovigilance services of the Galdakao-Usansolo Hospital and personel of the Pharmacy service of the Basurto Hospital; also, Oier Lopez de Lacalle, it was for the reading group he is conducting at IXA that we approached first to deep learning; moreover, Josu Goikoetxea for his advice and helpful discussions on word embeedings; and not least, the anonymous reviewers for their constructive criticism. This work was partially funded by the Spanish Ministry of Science and Innovation (EXTRECM: TIN2013-46616-C2-1-R, TADEEP: TIN2015-70214-P) and the Basque Government (DETEAMI: Department of Health 2014111003).

## References

Rodrigo Agerri and German Rigau. 2016. Robust multilingual named entity recognition with shallow semi-supervised features. *Artificial Intelligence*, 238:63 – 82.

Alfred V. Aho and Jeffrey D. Ullman. 1972. *The Theory of Parsing, Translation and Compiling*, volume 1. Prentice-Hall, Englewood Cliffs, NJ.

American Psychological Association. 1983. *Publications Manual*. American Psychological Association, Washington, DC.

Riza Batista-Navarro, Jennifer Hammock, William Ulate, and Sophia Ananiadou. 2016. A text mining framework for accelerating the semantic curation of literature. In *International Conference on Theory and Practice of Digital Libraries*, pages 459–462. Springer.

Antoine Bordes, Nicolas Usunier, Alberto Garcia-Duran, Jason Weston, and Oksana Yakhnenko. 2013. Translating embeddings for modeling multi-relational data. In *Advances in Neural Information Processing Systems*, pages 2787–2795.

Ashok K. Chandra, Dexter C. Kozen, and Larry J. Stockmeyer. 1981. Alternation. *Journal of the Association for Computing Machinery*, 28(1):114–133.

Kevin Bretonnel Cohen and Dina Demner-Fushman. 2014. *Biomedical Natural Language Processing*. Natural Language Processing. John Benjamins Publishing Company.

Hercules Dalianis, Aron Henriksson, Maria Kvist, Sumithra Velupillai, and Rebecka Weegar. 2015. Health bank - a workbench for data science applications in healthcar. In J. Krogstie, G. Juel-Skielse, and V. Kabilan, editors, *Proceedings of the CAiSE-2015 Industry Track co-located with 27th Conference on Advanced Information Systems Engineering*, volume 1381, pages 1–18.

Hercules Dalianis. 2014. Clinical text retrieval-an overview of basic building blocks and applications. In *Professional Search in the Modern World*, pages 147–165. Springer.

Lorenzo Ferrone and Fabio Massimo Zanzotto. 2013. Linear compositional distributional semantics and structural kernels. In *Joint Symposium on Semantic Processing.*, page 85. Citeseer.

Association for Computing Machinery. 1983. In *Computing Reviews*, volume 24, pages 503–512.

Dan Gusfield. 1997. *Algorithms on Strings, Trees and Sequences*. Cambridge University Press, Cambridge, UK.

Aron Henriksson, Jing Zhao, Henrik Bostrom, and Hercules Dalianis. 2015. Modeling electronic health records in ensembles of semantic spaces for adverse drug event detection. In *Bioinformatics and Biomedicine (BIBM), 2015 IEEE International Conference on*, pages 343–350. IEEE.

Aron Henriksson. 2015. Representing clinical notes for adverse drug event detection. In *Proceedings of the Sixth International Workshop on Health Text Mining and Information Analysis*, pages 152–158, Lisbon, Portugal, September. Association for Computational Linguistics.

Shoaib Jameel and Steven Schockaert. 2016. Entity embeddings with conceptual subspaces as a basis for plausible reasoning. In *European Conference on Artificial Intelligence*, volume 22, pages 1353–1361.

Markus Kreuzthaler and Stefan Schulz. 2015. Detection of sentence boundaries and abbreviations in clinical narratives. *BMC medical informatics and decision making*, 15(Suppl 2):S4.

Yann LeCun, Yoshua Bengio, and Geoffrey Hinton. 2015. Deep learning. *Nature*, 521(7553):436–444.

Tomas Mikolov, Ilya Sutskever, Kai Chen, Greg S Corrado, and Jeff Dean. 2013a. Distributed representations of words and phrases and their compositionality. In *Advances in neural information processing systems*, pages 3111–3119.

Tomas Mikolov, Wen-tau Yih, and Geoffrey Zweig. 2013b. Linguistic regularities in continuous space word representations. In *HLT-NAACL*, pages 746–751.

Tomas Mikolov. 2016. word2vec: Tool for computing continuous distributed representations of words. Accessed 2016-07-08, https://code.google.com/p/word2vec/.

Jeff Mitchell and Mirella Lapata. 2008. Vector-based models of semantic composition. In *ACL*, pages 236–244.

Maria Carolina Monard and Gustavo EAPA Batista. 2002. Learmng with skewed class distrihutions. *Advances in Logic, Artificial Intelligence, and Robotics: LAPTEC 2002*, 85:173.

Tingting Mu, Xinglong Wang, Jun'ichi Tsujii, and Sophia Ananiadou. 2010. Imbalanced classification using dictionary-based prototypes and hierarchical decision rules for entity sense disambiguation. In *Proceedings of the 23rd International Conference on Computational Linguistics: Posters*, pages 851–859. Association for Computational Linguistics.

Jonathan R. Nebeker, Paul Barach, and Matthew H. Samore. 2004. Clarifying adverse drug events: A clinician's guide to terminology, documentation, and reporting. *Annals of Internal Medicine*, 140(10):795–801.

Aurélie Névéol, Hercules Dalianis, Guergana Savo va, and Pierre Zweigenbaum. 2014. Panel: Clinical natural language processing in languages oth er than english. In *American Medical Informatics Association (AMIA)*.

Maximilian Nickel, Kevin Murphy, Volker Tresp, and Evgeniy Gabrilovich. 2016. A review of relational machine learning for knowledge graphs. *Proceedings of the IEEE*, 104(1):11–33.

Naoaki Okazaki, Sophia Ananiadou, and Jun'ichi Tsujii. 2010. Building a high-quality sense inventory for improved abbreviation disambiguation. *Bioinformatics*, 26(9):1246–1253.

Clifton Phua, Damminda Alahakoon, and Vincent Lee. 2004. Minority report in fraud detection: classification of skewed data. *Acm sigkdd explorations newsletter*, 6(1):50–59.

Sameer Pradhan, Noémie Elhadad, Wendy Chapman, Suresh Manandhar, and Guergana Savova. 2014. Semeval-2014 task 7: Analysis of clinical text. In *Proceedings of the 8th International Workshop on Semantic Evaluation (SemEval 2014)*, pages 54–62, Dublin, Ireland, August. Association for Computational Linguistics and Dublin City University.

Pontus Stenetorp, Sampo Pyysalo, Goran Topić, Tomoko Ohta, Sophia Ananiadou, and Jun'ichi Tsujii. 2012. Brat: a web-based tool for nlp-assisted text annotation. In *Proceedings of the Demonstrations at the 13th Conference of the European Chapter of the Association for Computational Linguistics*, pages 102–107. Association for Computational Linguistics.

Paul Thompson, Riza Theresa Batista-Navarro, Georgios Kontonatsios, Jacob Carter, Elizabeth Toon, John Mc-Naught, Carsten Timmermann, Michael Worboys, and Sophia Ananiadou. 2016. Text mining the history of medicine. *PloS one*, 11(1):e0144717.

Zhen Wang, Jianwen Zhang, Jianlin Feng, and Zheng Chen. 2014. Knowledge graph embedding by translating on hyperplanes. In *AAAI*, pages 1112–1119. Citeseer.

WHO. 2014. International classification of diseases (ICD). World Health Organization.

Fabio Massimo Zanzotto, Ioannis Korkontzelos, Francesca Fallucchi, and Suresh Manandhar. 2010. Estimating linear models for compositional distributional semantics. In *Proceedings of the 23rd International Conference on Computational Linguistics*, pages 1263–1271. Association for Computational Linguistics.

Congle Zhang, Stephen Soderland, and Daniel Weld. 2015. Exploiting parallel news streams for unsupervised event extraction. *Transactions of the Association for Computational Linguistics*, 3:117–129.

Deyu Zhou, Liangyu Chen, and Yulan He. 2015. An unsupervised framework of exploring events on twitter: Filtering, extraction and categorization.

# A Dataset for ICD-10 Coding of Death Certificates: Creation and Usage


**Thomas Lavergne**
LIMSI, CNRS, Univ. Paris-Sud
Université Paris-Saclay
F-91405 Orsay
`lavergne@limsi.fr`

**Aurélie Névéol**
LIMSI, CNRS
Université Paris-Saclay
F-91405 Orsay
`neveol@limsi.fr`

**Aude Robert**
CépiDC
80, rue du Général Leclerc
F-94276 le Kremlin-Bicêtre Cedex
`aude.robert@inserm.fr`

**Cyril Grouin**
LIMSI, CNRS
Université Paris-Saclay
F-91405 Orsay
`grouin@limsi.fr`

**Grégoire Rey**
CépiDC
80, rue du Général Leclerc
F-94276 le Kremlin-Bicêtre Cedex
`gregoire.rey@inserm.fr`

**Pierre Zweigenbaum**
LIMSI, CNRS
Université Paris-Saclay
F-91405 Orsay
`pz@limsi.fr`


## Abstract


Very few datasets have been released for the evaluation of diagnosis coding with the International Classification of Diseases, and only one so far in a language other than English. This paper describes a large-scale dataset prepared from French death certificates, and the problems which needed to be solved to turn it into a dataset suitable for the application of machine learning and natural language processing methods of ICD-10 coding. The dataset includes the free-text statements written by medical doctors, the associated meta-data, the human coder-assigned codes for each statement, as well as the statement segments which supported the coder's decision for each code. The dataset comprises 93,694 death certificates totalling 276,103 statements and 377,677 ICD-10 code assignments (3,457 unique codes). It was made available for an international automated coding shared task, which attracted five participating teams. An extended version of the dataset will be used in a new edition of the shared task.


## 1 Introduction

Over the past decade, biomedical named entity recognition (NER) and concept normalization have been widely covered in NLP challenges. Different types of texts were explored: clinical texts were used in the CMC (Pestian et al., 2007) and the i2b2 NLP Challenges (Uzuner et al., 2007; Uzuner et al., 2011) while the biomedical literature provided material for the BioNLP-Shared Tasks (Kim et al., 2011; Nédellec et al., 2015). Few challenges offered datasets in more than one languages, such as the CLEF ER (Rebholz-Schuhmann et al., 2013) and CLEF eHealth Challenges (Goeuriot et al., 2015)

The assignment of codes from the International Classification of Diseases (ICD) to clinical texts is primarily used for billing purposes but also has a wide range of applications including epidemiological studies (Woodfield et al., 2015), monitoring disease activity (Koopman et al., 2015a), or predicting cancer incidence through retrospective and prospective studies (Bedford et al., 2014). Nevertheless, useful results can only be achieved if ICD code assignment is accurate (Mieno et al., 2016), and studies evidenced that it is a challenging task even when performed manually (Dalianis, 2014).

This is a motivation for creating shareable datasets for ICD coding from natural language text: text corpora annotated with associated ICD codes that can be used to train and evaluate automatic coding systems. Automatic coding has the potential to reduce the cost of physician involvement in the coding process and to increase the consistency of coding.

A potential source of ICD coding datasets comes from death certificates, which are coded in countries around the world according to the World Health Organization (WHO) international standards, using ICD-10. This coding process exists in virtually every country, hence in a large variety of languages. We describe herein the creation of a large-scale ICD coding dataset from death certificates, instantiated in the case of France and the French language. This experience can pave the way for other instantiations in other countries and languages.






*Proceedings of the Fifth Workshop on Building and Evaluating Resources for Biomedical Text Mining (BioTxtM 2016),*
pages 60–69, Osaka, Japan, December 12th 2016.

We first review related work on ICD coding datasets, briefly mentioning associated automated coding methods. We then present the material we started from, the issues we encountered and how we solved them. We describe the resulting data and its use in an international shared task.

## 2   Related work

In 2007, the Computational Medicine Center (CMC) challenge proposed to identify ICD-9-CM disease codes on a corpus of outpatient chest x-ray and renal procedures (Pestian et al., 2007). In those documents, two sections are identified as more likely to yield codes: 'clinical history' and 'impression'. Both training set and test set are well balanced (respectively 978 and 976 documents). The corpus targeted a subset of only 45 ICD-9-CM codes so that each one of the 94 distinct combination of codes from the test set were seen during the training stage. The best system used a decision tree and achieved a 0.89 F-measure on the test set.

Apart from the CMC challenge, various studies have addressed automatic ICD-10 coding. Koopman et al. (2015a) classified Australian death certificates into 3-digit ICD-10 codes such as *E10* with SVM classifiers based on n-grams and SNOMED CT concepts, and with rules. They also trained SVM classifiers (Koopman et al., 2015b) to find ICD-10 diagnostic codes for death certificates. In contrast to the dataset presented here, they only addressed cancer-related certificates. In addition, they tackled the level of 3-digit ICD-10 codes (e.g., *C00, C97*) instead of the full 4-digit level usually required for ICD-10 coding (e.g., *C90.2*). Another important difference is that they focused on the underlying cause of death, i.e., one diagnosis per death certificate. The present dataset keeps all the diagnoses mentioned in each statement of a given death certificate, so that the number of codes to assign to a certificate varies from document to document and is not known *a-priori*. This dataset is intended to support a statement-coding task rather than as a certificate-coding task.

Perotte et al. (2014) took advantage of the presence of ICD-9 codes in the MIMIC-II database along with free text notes. They tested the use of the hierarchical structure of the ICD codes system to improve automatic coding. They compared two coding approaches to assign ICD-9 codes to documents, using SVM classifiers: one took into account the hierarchical structure of ICD-9 codes (hierarchy-based classifier); the other did not (flat classifier). They report higher recall (0.300) and F-measure (0.395) when using the hierarchy-based classifier.

All of this work addressed English language free text. Additionally, ICD-10 coding shared tasks from Japanese clinical records were organized at NTCIR-11 (MedNLP-2) (Aramaki et al., 2014) and NTCIR-12 (MedNLPDoc) (Aramaki et al., 2016). The latter included 200 medical records with an average 7.82 sentences and 3.86 ICD codes per record, totalling 552 distinct codes. However, the inter-annotator agreement was low, with an F-measure of 0.235. The best system obtained an F-measure of 0.348.

We present here the construction and use of a much larger-scale ICD-10 coding dataset in French. Instead of clinical records, it is based on much shorter narratives, viz. death certificates.

## 3   Material and methods

This section describes the original data; it presents the issues that prevented direct use for a shared task as well as the processing methods we designed to create a dataset suitable for a shared task.

### 3.1   The coding process at the French WHO collaborative center

Causes of death statistics are essential data to monitor population health, undertake epidemiological studies and international comparisons.

Death certification by a medical practitioner is a mandatory procedure for any death occuring on the French territory. It can be done on a paper certificate or through a secure Web application. In 2007, electronic certification was introduced in France with the objective (among others) to provide a much quicker process for health surveillance and alert systems (Pavillon et al., 2007). Currently, around 12% of death certificates are electronically certified. The system is run on a completely voluntary basis.

Paper death certificates are keyed in by contractors. In this process, contractors may normalize parts of the text to facilitate its subsequent coding; for instance, disease mentions may be replaced with an equivalent from a standard dictionary.

Causes of death data is centralized at the French Epidemiological Center for the Medical Causes of Death (CépiDc – Inserm). Death certificates are coded with the international software IRIS (Johansson and Pavillon, 2005) in order to assign a code selected from the International Classification of Diseases, tenth revision (ICD-10) to each reported nosologic entity. Then several ICD rules are applied in order to select the so-called underlying cause of death, which is used in most statistics compilations.

Death certificates are now increasingly produced electronically. While this makes the documents more easily available for machine processing, it also creates new challenges. Since electronic certificates are not handled by contractors, their variability of expression is higher than that of transcribed paper certificates; they can also contain spelling errors. Therefore, it is more difficult to handle automatic processing of electronic certificates compared to transcribed certificates which are currently handled by IRIS. This creates an additional motivation for testing state-of-the-art automatic coding methods on modern death certificates, as can be done in a shared task. For these reasons, we used electronic death certificates to create the dataset described herein.

## 3.2 Data produced by this coding process

In compliance with the World Health Organization (WHO) international standards (Wor, 2011), French death certificates are composed of two parts: Part I is dedicated to the reporting of diseases related to the main train of events leading directly to death, and Part II is dedicated to the reporting of contributory conditions not directly involved in the main death process. According to WHO recommendations, the completion of both parts is free of any automatic assistance that might influence the certifying physician.

In the course of coding practice, the data is stored in different files: a file that records the native text entered in the death certificates (called 'raw causes' thereafter) and a file containing the result of normalizing the text and assigning ICD codes (called 'computed causes' thereafter). An example of 'raw' and 'computed' causes is show below in Table 1.

## 3.3 Encountered issues

We found that the formatting of the data into raw and computed causes made it difficult to directly relate the codes assigned to original death certificate texts, which would reduce the interest of the data for a shared task. The main issues we identified were:

1. **Outside information needed.** Some coding decisions were made after complementary information was obtained through another channel, such as by contacting the author of the certificate. No trace of this communication is present inside the death certificate itself, hence its contents are not relevant as a source for coding.

2. **Alignment challenge.** The correspondence between the 'computed causes' records in the computed causes file and the statements in the raw causes file could not be easily recovered through the information present in these files. The raw causes file used actual line numbers of the source death certificate (1–4 and 5), but the computed causes file sometimes did not keep the order of the causes as mentioned in the raw causes, and used line numbering that could arbitrarily differ from that of the raw causes. Further more, the text of the computed causes consists of a normalized excerpt of the raw causes text that lead to the specific code assignment. In practice, this means that the specific text strings were related, but often not identical.

The certificates which needed outside information to assign the correct code could be identified through the mention of conditions that prevent a specific code assignment: *décès de cause inconnue* (*unknown cause of death*), *autopsie en cours* (*autopsy requested*) or through the automatic detection of incoherence between the cause mention and the patient age or gender. In those circumstances, a letter is addressed to the doctor, in order to request additional information. Every year, about 1,800 letters are sent and 500 answers received. With this feedback, codes are directly assigned to the corresponding certificates without revising the original text; instead, a free text comment reporting on the supporting correspondance is entered in the coding software. The certificates meeting these criteria were then removed from the dataset.

The alignment issue required to find a method to *align* the source statements with the computed causes records. We describe this method in the next section.

### 3.4 Pre-processing of death certificate through alignment

The goal of the alignment process is to obtain (statement, code) pairs, where the statement includes the original text and its associated meta-information, as per the raw causes file, and the code is one of those which should be assigned to this statement as per the computed causes file, together with the associated normalized text. Input statements with multiple codes are repeated in multiple (statement, code) pairs.

A sample document is presented in table 1. This example illustrates different types of difficulty of the alignment step:

- cause order is reversed (e.g., *choc septique* appears in line 1 of the raw causes but in line 3 of the computed causes),

- multiple causes are merged on a single raw line (e.g. *peritonite stercorale* and *perforation colique* on line 2),

- different capitalization and stopwords (e.g. see line 3 of aligned causes),

- different spelling. There is no occurrence of this in our sample document; however, a raw cause such as *bactériémie à K. pneumoniae* would be normalized to *bactériémie klebsiella pneumoniae*, using a variant of the name of the bacteria involved in the reported infection.

Table 1: A sample document from the CépiDC French Death Certificates Corpus: alignment of the raw causes and computed causes. English translations for each text line are provided in footnotes.

| data type | line | text | normalized text | ICD codes |
|---|---|---|---|---|
| **Raw causes** | 1 | choc septique[1] | | - |
| | 2 | peritonite stercorale sur perforation colique[2] | | - |
| | 3 | Syndrome de détresse respiratoire aiguë[3] | | - |
| | 4 | defaillance multivicerale[4] | | - |
| | 5 | HTA[5] | | - |
| **Computed causes** | 1 | | defaillance multivicerale | R57.9 |
| | 2 | | syndrome détresse respiratoire aiguë | J80.0 |
| | 3 | | choc septique | A41.9 |
| | 4 | | peritonite stercorale | K65.9 |
| | 5 | | perforation colique | K63.1 |
| | 6 | | hta | I10.0 |
| **Aligned causes** | 1 | choc septique | choc septique | A41.9 |
| | 2 | peritonite stercorale sur perforation colique | peritonite stercorale | K65.9 |
| | 2 | peritonite stercorale sur perforation colique | perforation colique | K63.1 |
| | 3 | Syndrome de détresse respiratoire aiguë | syndrome détresse respiratoire aiguë | J80.0 |
| | 4 | defaillance multivicerale | défaillance multiviscérale | R57.9 |
| | 5 | HTA | hta | I10.0 |

Our alignment method relied both on the order that causes and codes occurred in the files and on string similarity between the texts of raw and computed causes. More specifically, the principles we followed to reconcile raw and computed causes were the following:

---

[1] *septic shock*
[2] *colon perforation leading to stercoral peritonitis*
[3] *Acute Respiratory Distress Syndrome*
[4] *multiple organ failure*
[5] *HBP: High Blood Pressure*

- Alignments have the form $(0, 1) \rightarrow m$

- All computed causes must be supported by an input statement

- No alignment should have the form $n \rightarrow m$. However, some death certificates contain separate input statements which must be taken as a whole to produce a relevant code. An example is the set of two lines *1. Strangulation au lien (ligature strangulation)* and *2. Suicide* which is coded as X70.9 (*Intentional self-harm by hanging strangulation and suffocation home during unspecified activity*), where the suicide must be coded by taking into account the specific circumstance that lead to it (here, strangulation). In such cases we kept the input statements separate. The most generic statement (e.g. *suicide*) was considered inconclusive and did not receive a code assignment while the 'head' statement (e.g. *ligature strangulation*, which provided the defining information for code assignment) was aligned with the output code.

To align the statements, we used a model originally intended for bilingual word alignment in parallel sentences: a log-linear reparameterization of the IBM2 model (Dyer et al., 2013). The alignments were produced from the computed clauses without allowing for null alignment in order to satisfy our constraints, and with a Dirichlet prior to favor diagonal alignments.

The model underperforms on multi-word segments as it relies on co-occurrence counts of raw and computed causes, which are very sparse. To overcome this problem, both causes were pre-processed by removing stopwords and applying stemming. Next, the Damerau-Levenshtein distance between two segments was linearly combined with the occurrence count to act as a prior on the alignment probabilities.

## 4   Results

We applied the above-described methods to the 2006–2013 death certificates created by the electronic work-flow and describe the resulting data and its usage.

### 4.1   Corpus characteristics

Table 2 presents the fields found in each line of the produced dataset. One line is produced for each (input line, output code) pair. Some input lines have no associated output code: the corresponding values are empty. As explained in Section 3.4, this also occurs when two "raw cause" input lines need to be considered together to be coded. In that case, only one of them has an associated code.

The dataset was split into training and test sets: the training set contains statements of years 2006–2012, and the test set contains statements of year 2013. We now provide more detail on the training set.

The distribution of statement length in tokens, after stop-word removal (French stop words of the NLTK toolkit), is shown on Figure 1a. It shows that statement length follows a Zipfian distribution from length 2 to length 31. Statements over 20 tokens are rare (455 = 0.17%), over 10 tokens too (9538 = 3.6%). The maximum length of a statement is 120 tokens.

Figure 1b shows the most frequent codes. The top five are R092 (Respiratory arrest), A419 (Septicaemia, unspecified), R688 (Other specified general symptoms and signs), I10 (Essential (primary) hypertension), I509 (Heart failure, unspecified). These top diagnoses, as well as those in the rest of the figure, display a mixture of very general diagnoses (*unspecified*, *other*) and most frequent causes of death (infection, hypertension, pneumonia, cancer, etc.).

ICD-10 is divided into 21 chapters. Figure 1c shows the number of codes in each chapter in the training set. The most represented chapters are Chapters IX (codes I00–I999, Diseases of the circulatory system), II (C00–D489, Neoplasms), XVIII (R00–R999, Symptoms, signs and abnormal clinical and laboratory findings, not elsewhere classified), etc. Figure 1d shows the number of occurrences of each age group for each chapter. A few chapters have a skewed distribution of age groups: P00-P969 (Certain conditions originating in the perinatal period: 99.8% for age group 0), Q00-Q999 (Congenital malformations, deformations and chromosomal abnormalities, 55.4% for age group 0), and O00-O999 (Pregnancy, childbirth and the puerperium: 15.4% for age group 25, 46.3% for 30, 34.6% for 35).

| Field | Contents |
|---|---|
| DocID | Death certificate ID. |
| YearCoded | Year the death certificate was processed by the human coder. |
| Gender | gender of the deceased |
| Age | Age at the time of death, rounded to the nearest five-year age group. |
| LocationOfDeath | Location of death, according to the following categories: 1 = Home; 2 = Hospital; 3 = Private Clinic; 4 = Hopice, Retirement home; 5 = Public place; 6 = Other Location. |
| LineID | Line number within the death certificate. Note that if a statement is assigned multiple ICD10 codes, it is repeated for each code, each time with the same LineID. |
| RawText | Raw text entered in the death certificate. |
| IntType | Type of time interval the patient had been suffering from coded cause, according to the following categories: 1 = minutes; 2 = hours; 3 = days; 4 = months; 5 = years. |
| IntValue | Length of time the patient had been suffering from coded cause; for example, if the patient had been experiencing the cause for 6 months, *IntValue* should be 6 and *IntType* should be 4. |
| CauseRank | Rank of the ICD10 code assigned by the human coder. The rank (e.g., *2-1*) is composed of two items found in the original CausesCalculees file: the number of the line (NumLigne, e.g., *2*) followed by the rank of the cause in that line (RangCause, e.g., *1*). |
| StandardText | Dictionary entry or excerpt of the raw text that supports the selection of an ICD10 code. |
| ICD10 | Gold standard ICD10 code. |

Table 2: Fields in each row of the dataset. The last three fields are the output of the coding process.

## 4.2 Use in a shared task

The resulting dataset was used in an international shared task (Névéol et al., 2016). The certificates corresponding to year 2006-2012 were used as a training set (N=65,844) while certificates corresponding to the year 2013 were used as a test set (N=27,850). A small number of codes (N=244, about 10% of the unique codes in the test set) in the test set were unseen in the training set. Five teams from three countries submitted a total of seven runs for this task. Participating teams used methods relying either on knowledge-base linking or statistical machine learning. Table 3 shows the performance of the official runs, compared to a baseline run, which consisted in assigning codes to lines in the test set when an exactly identical line was also found in the training set. When the line occurred multiple times in the training set, the most frequent code was selected. It can be seen from the table that all runs submitted by participants outperformed the baseline by at least 20 points in F-measure, thus demonstrating that the state of the art in ICD10 coding is quite advanced.

We examined the relative difficulty of finding each expected statement code for the submitted systems: for each death certificate statement and expected code for this statement, we counted the number of systems which correctly found this code. Figure 2 shows the results.

We found out that among the 110767 distinct entries of the test dataset, 29100 were easy to find: all systems found the correct answer; 25215 were fairly easy: all but one system found them; 20743 were less easy (3 systems); 15933 were harder (2 systems); 10685 were rather hard: only one system found them; and 7714 were hard: no system found them at all. The latter may help identify to difficult, hence interesting problems, such as codes which need to refer to the broader context of the full death certificate, beyond the current individual statement, to be assigned properly. They may also point at cases where human coding might not be correct.

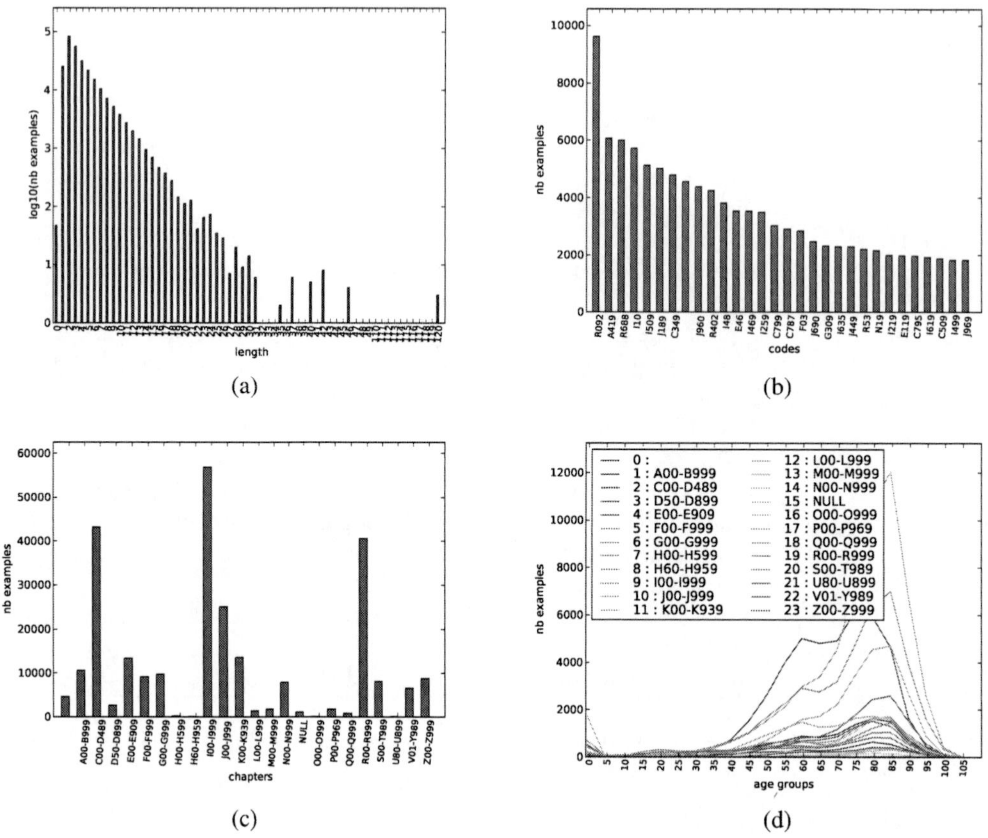

Figure 1: Statistics on training set. (a) Distribution of statement length (after normalization, log-y scale). (b) Most frequent codes. (c) Distribution of ICD-10 chapters. (d) Occurrences of age groups for each chapter.

## 5 Discussion

With 93,694 death certificates totalling 276,103 statements and 377,677 ICD-10 code assignments (3,457 distinct codes), the size of the presented dataset is comparable to the largest so far on English (Perotte et al., 2014) (22,815 discharge summaries and 215,826 ICD9 codes (5,030 distinct codes)), and is several orders of magnitude above the other ICD coding datasets we identified (Pestian et al., 2007; Aramaki et al., 2014; Aramaki et al., 2016).

An important difference though is that the present dataset consists of death certificate statements, whereas the other cited datasets are made of clinical records such as discharge summaries. Death certificate statements are fairly short and focused on nosologic entities, whereas clinical records are usually longer and mention a broader set of entities and events. Medical records exhibit a large range of sizes however: for instance, texts in the MedNLPDoc dataset (Aramaki et al., 2016) contained on average 7.82 sentences.

A consequence of the difference between death certificate statements and for instance discharge summaries is that death certificate statement coding might be more easily addressed as a text classification task, whereas clinical record coding may need to rely on a step of entity detection and normalization methods to identify more relevant pieces of information before ICD coding proper. This makes the clinical record coding task more difficult and explains the lower F-measures obtained in that context.

Future plans include the extension of the present dataset with death certificates of more recent years

Table 3: System performance for ICD10 coding on the death certificate test corpus. A * symbol indicates statistically significant difference of a run with the runs ranked before and after it, according to a Student test.

| Team | TP | FP | FN | Precision | Recall | F-measure |
|---|---|---|---|---|---|---|
| TeamA-run2* | 88497 | 11423 | 20321 | 0.886 | **0.813** | **0.848** |
| TeamA-run1* | 87404 | 10823 | 21414 | **0.890** | 0.803 | 0.844 |
| TeamB-run2* | 71319 | 9479 | 37499 | 0.882 | 0.655 | 0.752 |
| TeamB-run1* | 66954 | 15605 | 41864 | 0.811 | 0.615 | 0.700 |
| TeamC-run1* | 72192 | 31480 | 36626 | 0.696 | 0.663 | 0.680 |
| TeamD-run1* | 61874 | 19002 | 46984 | 0.765 | 0.569 | 0.652 |
| TeamE-run1* | 57256 | 40650 | 51562 | 0.585 | 0.526 | 0.554 |
| Baseline-Zipf-Top1* | 26688 | 23610 | 82130 | 0.531 | 0.245 | 0.336 |

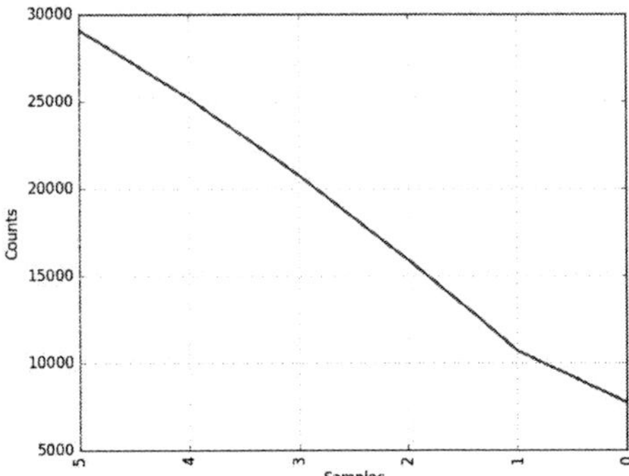

Figure 2: Distribution of coding difficulty based on system results, based on the best run of each of the five participating teams. Samples = number of systems which found the expected code for a statement. Counts = number of (statement, code) pairs found by a given number of systems.

as they are processed by human coders. In an upcoming edition of the shared task, he dataset described herein will be used as a training set while more recent data will be offered as a test set. This time-ordered distribution of certificates in the datasets is guided by the practical use case of coding death certificates, where historical data is available to coders who then need to work with current data. The goal of this series of shared tasks is to engage the community in the development of ICD-10 coding methods that can then be integrated to coders work flow as coding assistance and productivity enhancing tools.

We also plan to include additional languages in future datasets, as other WHO collaborating centers express their interest in this enterprise. We hope that the development of a multilingual ICD-10 coding dataset will foster the development of portable methods that can be easily adapted to several languages.

## 6  Conclusion

This paper presents a new dataset for ICD-10 coding based on death certificates in French. This is a large dataset comprising death certificate statements in a language other than English as well as rich metadata and professionally assigned gold-standard ICD10 codes. The preparation of the dataset involved the use of complex alignment techniques to ensure the quality of the text-code pairings. It was shown to be a suitable tool for evaluating the state of the art in ICD-10 coding in an international shared task. In future work we plan to enhance the dataset with newer data for French as well as other languages in order to foster global approaches to ICD10 coding.

# 7 Acknowledgements

This work was partially funded by the European Union's Horizon 2020 Marie Sklodowska Curie Innovative Training Networks—European Joint doctorate (ITN-EJD) under grant agreement No:676207, Methods in Research on Research (MiRoR).

This work was supported in part by the French National Agency for Research under the CABeRneT[6] ANR-13-JS02-0009-01 grant.

## References

Eiji Aramaki, Mizuki Morita, Yoshinobu Kano, and Tomoko Ohkuma. 2014. Overview of the NTCIR-11 MedNLP-2 task. In *Proceedings of the 11th NTCIR Conference*, Tokyo Japan.

Eiji Aramaki, Mizuki Morita, Yoshinobu Kano, and Tomoko Ohkuma. 2016. Overview of the NTCIR-12 MedNLPDoc task. In *Proceedings of the 12th NTCIR Conference on Evaluation of Information Access Technologies*, Tokyo Japan.

Ronald L. Bedford, Spencer G. Lourens, Charles F. Lynch, Brian J. Smith, and R. William Field. 2014. Utility of death certificate data in predicting cancer incidence. *Am J Ind Med*, 57(2):153–62.

Hercules Dalianis. 2014. Clinical text retrieval - an overview of basic building blocks and applications. In Georgios Paltoglou, Fernando Loizides, and Preben Hansen, editors, *Professional Search in the Modern World: COST Action IC1002 on Multilingual and Multifaceted Interactive Information Access*, pages 147–165. Springer International Publishing, Cham.

Chris Dyer, Victor Chahuneau, and A. Noah Smith. 2013. A simple, fast, and effective reparameterization of ibm model 2. In *Proceedings of the 2013 Conference of the North American Chapter of the Association for Computational Linguistics: Human Language Technologies*, pages 644–648. Association for Computational Linguistics.

Lorraine Goeuriot, Liadh Kelly, Hanna Suominen, Leif Hanlen, Aurélie Névéol, Cyril Grouin, João Palotti, and Guido Zuccon, 2015. *Overview of the CLEF eHealth Evaluation Lab 2015*, pages 429–443. Springer International Publishing, Cham.

Lars Age Johansson and Gérard Pavillon. 2005. IRIS: A language-independent coding system based on the NCHS system MMDS. In *WHO-FIC Network Meeting*, Tokyo, Japan.

Jin-Dong Kim, Sampo Pyysalo, Tomoko Ohta, Robert Bossy, Ngan Nguyen, and Jun'ichi Tsujii. 2011. Overview of BioNLP shared task 2011. In *Proceedings of BioNLP Shared Task 2011 Workshop*, pages 1–6, Portland, Oregon, USA, June. Association for Computational Linguistics.

Bevan Koopman, Sarvnaz Karimi, Anthony Nguyen, Rhydwyn McGuire, David Muscatello, Madonna Kemp, Donna Truran, Ming Zhang, and Sarah Thackway. 2015a. Automatic classification of diseases from free-text death certificates for real-time surveillance. *BMC Med Inform Decis Mak*, 15:53.

Bevan Koopman, Guido Zuccon, Anthony Nguyen, Anton Bergheim, and Narelle Grayson. 2015b. Automatic ICD-10 classification of cancers from free-text death certificates. *Int J Med Inform*, 84(11):956–965, November.

Makiko Naka Mieno, Noriko Tanaka, Tomio Arai, Takuya Kawahara, Aya Kuchiba, Shizukiyo Ishikawa, and Motoji Sawabe. 2016. Accuracy of death certificates and assessment of factors for misclassification of underlying cause of death. *J Epidemiol*, 26(4):191–8.

Claire Nédellec, Jin-Dong Kim, Sampo Pyysalo, Sophia Ananiadou, and Pierre Zweigenbaum. 2015. BioNLP Shared Task 2013: Part 1. *BMC Bioinformatics*, 16(Suppl 10), July.

A Névéol, KB Cohen, C Grouin, T Hamon, T Lavergne, L Kelly, L Goeuriot, G Rey, A Robert, X Tannier, and P Zweigenbaum. 2016. Clinical information extraction at the CLEF eHealth evaluation lab 2016. In *CLEF 2016 Online Working Notes*, pages 28–42. CEUR-WS.

Gérard Pavillon, Patrick Coilland, and Éric Jougla. 2007. Mise en place de la certification électronique des causes médicales de décès en France : premier bilan et perspectives [Implementation of the electronic certification of medical causes of death in France: first results and propects]. *Bulletin épidémiologique hebdomadaire*, 35-36:306–308, Sep 18.

---

[6]CABeRneT: Compréhension Automatique de Textes Biomédicaux pour la Recherche Translationnelle

Adler Perotte, Rimma Pivovarov, Karthik Natarajan, Nicole Weiskopf, Frank Wood, and Noémie Elhadad. 2014. Diagnosis code assignment: models and evaluation metrics. *J Am Med Inform Assoc*, 21(2):231–7.

John P. Pestian, Chris Brew, Pawel Matykiewicz, DJ Hovermale, Neil Johnson, K. Bretonnel Cohen, and Wlodzislaw Duch. 2007. A shared task involving multi-label classification of clinical free text. In *Biological, translational, and clinical language processing*, pages 97–104, Prague, Czech Republic, June. Association for Computational Linguistics.

Dietrich Rebholz-Schuhmann, Simon Clematide, Fabio Rinaldi, Senay Kafkas, ErikM. van Mulligen, Chinh Bui, Johannes Hellrich, Ian Lewin, David Milward, Michael Poprat, Antonio Jimeno-Yepes, Udo Hahn, and JanA. Kors. 2013. Entity recognition in parallel multi-lingual biomedical corpora: The CLEF-ER laboratory overview. In Pamela Forner, Henning Müller, Roberto Paredes, Paolo Rosso, and Benno Stein, editors, *Information Access Evaluation. Multilinguality, Multimodality, and Visualization*, volume 8138 of *Lecture Notes in Computer Science*, pages 353–367. Springer Berlin Heidelberg.

Özlem Uzuner, Yuan Luo, and Peter Szolovits. 2007. Evaluating the state-of-the-art in automatic de-identification. *J Am Med Inform Assoc*, 14:550–563.

Özlem Uzuner, Brett R. South, Shuying Shen, and Scott L DuVall. 2011. 2010 i2b2/VA challenge on concepts, assertions, and relations in clinical text. *J Am Med Inform Assoc*, 18(5):552–556, Sep-Oct. Epub 2011 Jun 16.

Rebecca Woodfield, Ian Grant, UK Biobank Stroke Outcomes Group, UK Biobank Follow-Up and Outcomes Working Group, and Cathie L. M. Sudlow. 2015. Accuracy of electronic health record data for identifying stroke cases in large-scale epidemiological studies: A systematic review from the UK biobank stroke outcomes group. *PLoS One*, 10(10):e0140533.

World Health Organization, 2011. *ICD-10. International Statistical Classification of Diseases and Related Health Problems. 10th Revision. Volume 2. Instruction manual.*

# A Corpus of Tables in Full-Text Biomedical Research Publications

**Tatyana Shmanina**[1,3], **Ingrid Zukerman**[1], **Ai Lee Cheam**[1], **Thomas Bochynek**[2,3], **Lawrence Cavedon**[4]

[1]Clayton School of Information Technology, Monash University, Australia
[2]Caulfield School of Information Technology, Monash University, Australia
[3]Data61, CSIRO, Melbourne, Australia
[4]School of Science, RMIT University, Australia
[1,2]`firstname.lastname@monash.edu`, [4]`firstname.lastname@rmit.edu.au`


## Abstract


The development of text mining techniques for biomedical research literature has received increased attention in recent times. However, most of these techniques focus on prose, while much important biomedical data reside in tables. In this paper, we present a corpus created to serve as a gold standard for the development and evaluation of techniques for the automatic extraction of information from biomedical tables. We describe the guidelines used for corpus annotation and the manner in which they were developed. The high inter-annotator agreement achieved on the corpus, and the generic nature of our annotation approach, suggest that the developed guidelines can serve as a general framework for table annotation in biomedical and other scientific domains. The annotated corpus and the guidelines are available at `http://www.csse.monash.edu.au/research/umnl/data/index.shtml`.


## 1 Introduction

Biomedical science generates vast quantities of data, which reside in publicly available databases and repositories of structured biomedical information, such as the Catalogue of Somatic Mutations in Cancer (Bamford et al., 2004) and the International Society for Gastrointestinal Hereditary Tumours Database (Plazzer et al., 2013). In order to be useful to researchers, data sources must contain precise and reliable information, and therefore are typically manually curated by biomedical professionals (Campos et al., 2013), which leads to a "curation bottleneck". As a result, automatic information extraction from biomedical literature has become an important task.

The development of approaches for automatic and semi-automatic information extraction requires annotated corpora for training and evaluating text mining systems. To date, biomedical text mining has focused on extracting information from prose, yielding a wealth of diverse annotated corpora for unstructured text. For example, gold and silver standard corpora have been developed for a variety of tasks, such as named entity recognition (Doğan et al., 2014; Kim et al., 2003; Rebholz-Schuhmann et al., 2010; Verspoor et al., 2013), entity linking (Bada et al., 2012; Doğan et al., 2014), and relation and event extraction (Kim et al., 2003; Lee et al., 2016; Rosario and Hearst, 2004; Verspoor et al., 2013). The source of these datasets also varies, e.g., corpora comprising research abstracts (Doğan et al., 2014; Kim et al., 2003; Rebholz-Schuhmann et al., 2010) versus full-text journal articles (Bada et al., 2012; Lee et al., 2016; Verspoor et al., 2013).

In addition to prose, biomedical literature frequently presents information in other forms, such as tables and graphs. Several studies have shown that tables often contain important data and experimental results that are not mentioned in the main text of publications (Jimeno Yepes and Verspoor, 2013; Wong et al., 2009). At the same time, Jimeno Yepes and Verspoor (2013) have shown that text mining techniques developed for prose tend to under-perform when applied to tables, because of the difference in how information is presented in tables and in text. For example, the arrangement of cells, which is meaningful for understanding table contents, is not taken into account by classical prose mining techniques. This calls for the development of specialised approaches to information extraction from tables (*Table IE*) in






*Proceedings of the Fifth Workshop on Building and Evaluating Resources for Biomedical Text Mining (BioTxtM 2016)*,
pages 70–79, Osaka, Japan, December 12th 2016.

the biomedical literature. Advances in Table IE have been made in general and Web domains (Cafarella et al., 2008; Hignette et al., 2009; Hurst, 2000; Jannach et al., 2009; Limaye et al., 2010; Mulwad et al., 2013; Quercini and Reynaud, 2013; Van Assem et al., 2010; Venetis et al., 2011; Wang et al., 2012; Yakout et al., 2012; Yin et al., 2011; Yosef et al., 2011). However, Table IE has received comparatively little attention from the biomedical text mining community, which may be partly attributed to the limited availability of suitable annotated corpora.

In this paper, we introduce a new corpus of tables obtained from full-text biomedical research papers in two areas of genetics: human cancer and mouse — the tables in the human cancer papers cover topics such as genetic aberrations and patient and tumour characteristics; and the tables in the mouse papers include distributions of genotypes and phenotypes, and parameters and outcomes of genetic analyses. This corpus was developed to support our work in Table IE, which focuses on relation extraction and fine-grained named entity recognition. The tables in our corpus were supplemented with the following annotations: (1) concepts in table cells; (2) classification of table cells into homogeneous cell groups; (3) fine-grained cell types of each homogeneous cell group; and (4) relations between cell groups.

In Section 2, we motivate the design of the corpus, and describe the created corpus, the annotation schema and the annotation guidelines. Section 3 details the corpus construction process. The characteristics of the developed corpus are discussed in Section 4, followed by concluding remarks in Section 5.

## 2 Corpus Design

The design of our corpus and associated annotation schemas is closely aligned with the requirements of our project on information extraction from tables in biomedical research papers. However, both the corpus and the schemas are general enough to be of value to the broader biomedical text mining community.

The presented corpus is the gold standard for two information extraction tasks: (1) mapping of table cells into fine-grained entity types, and (2) identification of relations between table cells. Both fine-grained entity types and relations are drawn from a domain vocabulary. The design of the corpus was strongly influenced by the characteristics of the biomedical tables we encountered, which have a variety of structures, and tend to be more complex than the structures typically considered by researchers in information extraction. For example, Limaye et al. (2010) and Mulwad et al. (2013), who worked on a table information extraction task similar to ours, but for a Web domain, assumed that tables have lattice-like structures, and that the objective of information extraction is to identify table column types and relations between columns. However, even for simple lattice-like biomedical tables, we often found that interesting relations could be built between columns and their headers, and between the headers themselves. For instance, the relation *associated_with* can be built between the header "Diet-induced Obesity" and the data cells "S [8]", "R [8]", "S [41]", "R [44]" and "S [18]" in the table in Figure 1b.

This motivated us to view each table as a collection of homogeneous groups of cells, rather than a collection of columns. We assume that (1) all cells within each homogeneous group of cells share the same fine-grained type, and (2) it is possible to define relations among cell groups that hold for each corresponding pair of cells inside the groups. These assumptions motivated the creation of four types of annotation: (1) *cell group*, which splits each table into sets of homogeneous cell groups; (2) *cell*

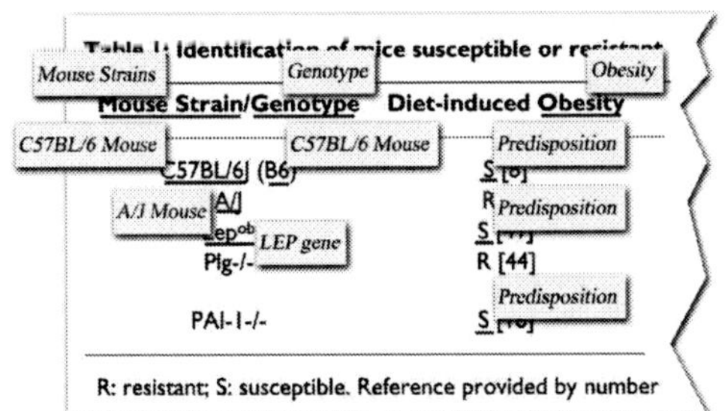

(a) Concept annotations. The annotated text spans are underlined, concept annotations are pictured in boxes.

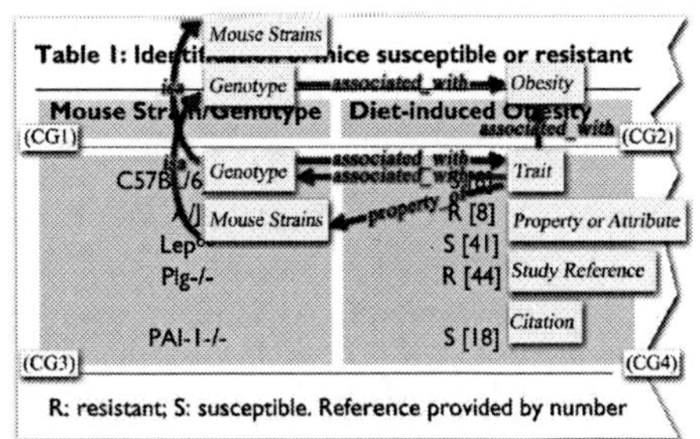

(b) Cell group, cell type and relation annotations. Cell groups (CG1)–(CG4) are highlighted, cell types are pictured in boxes, relations are represented as arrows.

Figure 1. Annotation example for a sample biomedical table from (Hoover-Plow et al., 2006)

*type*, which represents the mapping of all cells in a homogeneous group into a single fine-grained named entity label; (3) *concept*, which represents the mapping of utterances inside table cells (i.e., the syntactic heads of the utterances expanded with their modifiers) into their semantic equivalents from a domain vocabulary; and (4) *relation*, which represents relations between cell groups. Figure 1 illustrates a table annotated with these types of information.

## 2.1 Annotation Schema

**Cell Group Annotation.**  Each set of homogeneous cells in a table is assigned a unique identifier to distinguish between different cell groups.

**Concept, Cell Type and Relation Annotation.**  To generate fine-grained entity and relation labels, we used the National Cancer Institute (*NCI*) subset (Sioutos et al., 2007) of the Unified Medical Language System® (UMLS®) Metathesaurus (*UMLS-NCI*) and the UMLS Semantic Network (*UMLS SN*) as the basis for our annotation schema (UMLS release 2015AA). To illustrate, using this annotation schema, the text spans "Obesity" and "S" in Figure 1a are mapped to the UMLS-NCI concepts *Obesity* and *Predisposition* respectively to create concept annotations. The single-cell group "Diet-induced Obesity" is assigned the fine-grained cell type *Obesity* (Figure 1b) and the coarser cell type *Disease or Syndrome* from the UMLS SN. Finally, the relation *associated_with* from the UMLS SN is built between the cell group "Diet-induced Obesity" and the cell group comprising the data cells "S [8]", "R [8]", "S [41]", "R [44]" and "S [18]" (Figure 1b). No relations from UMLS-NCI can be built between these cell groups.

We chose the UMLS because of the lexical and information extraction tools for unstructured text that are distributed with the UMLS. We decided to use one subset of the UMLS Metathesaurus, because (1) the reduced size of the annotation schema reduces the complexity of the annotation tasks; and (2) the UMLS combines over 100 source vocabularies, and does not resolve conflicts among these vocabularies.

The **UMLS-NCI** subset was chosen because it is a large, comprehensive and heterogeneous controlled vocabulary, which focuses on genetics. It comprises over 110,000 biomedical Concept Unique Identifiers (*CUIs*) (e.g., *C0028754* for *Obesity* and *C0220898* for *Predisposition*), and more than one million relationships between concepts, drawn from 208 unique relation types (e.g., the generic hierarchical relations *parent-child*, and many specialised relations, such as *is_grade_of_disease* and *gene_mapped_to_disease*), thus providing a large set of fine-grained entity and relation labels. The concepts and relations from UMLS-NCI were used as the primary set of labels for concept, cell-type and relation annotation. However, despite its extensive scope, the UMLS-NCI's coverage of relations between the concepts in our dataset was very sparse. Specifically, it yielded only 222 relations in total for the entire corpus. We therefore expanded the set of relations in the UMLS-NCI schema with SRs from the UMLS SN.

The **UMLS SN** provides an alternative set of labels for cell types and relations that is smaller, and hence of lower granularity, than the labels in UMLS-NCI. It consists of (1) a set of Semantic Types (*STs*) that provide a broad subject categorisation of the concepts represented in the UMLS-NCI; and (2) a set of Semantic Relations (*SRs*) that can hold between STs. The UMLS SN contains 127 STs (e.g., *Organism Attribute* for the concepts *Age* and *Gender*, *Clinical Attribute* for the concepts *Tumour Stage* and *Cellular Differentiation*) and 54 SRs (e.g., *isa, causes, consists_of, interacts_with, assesses_effect_of* and *location_of*). Each concept in UMLS-NCI is assigned one or more STs;[1] and SRs defined between the STs in UMLS SN may or may not hold between particular concepts assigned to these STs. The incorporation of UMLS SN into our schema enabled the creation of 1625 additional relation annotations.

## 3   Corpus Construction

The construction of the dataset involved the following activities, described below: (1) document selection, (2) development of annotation guidelines, (3) document pre-processing and choice of distribution format, (4) annotation tool configuration and development, and (5) actual annotation process.

## 3.1   Document Selection

The following criteria were applied to select documents for our corpus: (1) the documents must represent full-text biomedical research articles containing at least one table; (2) the corpus must be diverse with

---

[1]When a concept was linked to several STs, our annotation guidelines required the exclusion of STs that were irrelevant in the context of the source table.

respect to the structure of tables, and representative of their distribution in biomedical research publications, in order to eliminate selection bias based on table structure; (3) the documents must be available in a structured format, preferably XML, to avoid the need to programmatically determine document and table structure; and (4) the articles must be available under non-restrictive licensing terms to enable future public release of our dataset. Finally, we preferred articles that were already included in other corpora with existing concept, named entity or relation annotations of unstructured text, in order to facilitate the future development of text mining tools for the joint analysis of free text and tabular content.

The application of these criteria resulted in the inclusion of papers from the following datasets:[2]

1. CRAFT Corpus (Bada et al., 2012). A subset of the CRAFT Corpus comprising 24 papers (50 tables) was included in our dataset. The CRAFT dataset, which comprises articles drawn from the Open Access subset of PubMed Central, is heterogeneous with respect to the content of the papers, and covers topics related to mouse genetics. The CRAFT dataset contains a mapping of concepts that appear in its free-text parts into seven open biomedical ontologies.

2. The Human Variome Project (HVP) Corpus (Verspoor et al., 2013). Nine out of ten papers (28 tables) from the HVP Corpus were included in our dataset. This corpus covers topics related to the genetics of human colon cancer. The free-text parts of the papers are annotated using a small annotation schema comprising eleven named entity classes and thirteen binary relations between the entity classes.

3. An additional subset of ten papers (22 tables) was randomly sampled from Open Access subsets of three datasets comprising papers about genomic variation (Jimeno Yepes and Verspoor, 2013; Wong et al., 2009). These datasets did not contain annotations of unstructured text, but several of these papers were previously used in (Shmanina et al., 2014).

### 3.2 Development of Annotation Guidelines

The table annotation guidelines were developed by the first author, who is a researcher in biomedical text mining. The guidelines contain four parts, each corresponding to a single annotation task: (1) cell group, (2) concept, (3) cell type, and (4) relation annotation.

The initial versions of the guidelines were developed through several iterative attempts to annotate ten tables using the guidelines. After each annotation iteration, we tested whether the strict application of the guidelines yielded consistent and objective annotations, and revised the guidelines as necessary. The final version of the guidelines was 49 pages long.

**Cell Group Annotation.**  Cells were collated into homogeneous cell groups according to two main guidelines: (1) every header cell should form its own cell group, and (2) data cells should be merged into maximum-size cell groups.

**Concept Annotation.**  These annotations were created due to the potential subjectivity of assigning a cell type to a cell group. For example, a cell group comprising the entries "C57/LJ", "AKR/J" and "NZB/BiNJ" can be potentially assigned UMLS-NCI concepts *[C0026809] Mice*, *[C1518614] Organism Strain* or *[C2985604] Biologic Entity Group*. By first annotating concepts for each table cell, it was relatively straightforward to derive cell type annotations. For example, the mentions above can be mapped into the concepts *[C1511387] C57L/J Mouse*, *[C1515841] AKR/J Mouse* and *[C1513862] NZB/BlNJ Mouse* — all of which have a common parent *[C0025927] Inbred Mouse Strains* in the UMLS-NCI concept hierarchy, which is then chosen to represent the cell type.

We based our concept annotation guidelines on those used in the CRAFT Corpus, which resulted in high inter-annotator agreement for concept annotations of free text (Bada et al., 2012). The main characteristics of these guidelines are: (1) a text is mapped into a concept from a vocabulary only if the concept is an exact semantic match for the text; and (2) the rules for the identification of text segments are syntax-based, and specify how to annotate nouns and noun phrases, adjectival and prepositional phrases, nested and overlapping mentions, etc.

We slightly modified the original CRAFT guidelines to better suit our table annotation task. Firstly, for each table cell, we annotated only the syntactic head of an utterance, expanded with as many of

---

[2] All the articles in our table dataset belong to the Open Access subset of PubMed Central.

its modifiers as possible. For example, given the table entry "Fragment length", UMLS-NCI contains concepts that are semantically equivalent to "Fragment" and "length", but there is no concept that is semantically equivalent to "Fragment length". Therefore, only "length", which is the syntactic head of the phrase, receives a concept annotation, and its modifier is excluded from the annotation. Secondly, the text inside table cells tends to be more concise than unstructured text. For instance, in a column listing colon cancer stages, such as "A", "B" and "C", "B" could stand for "stage B" or "Dukes B rectal cancer". We addressed this problem by stipulating that if it was not possible to annotate a concept using the guidelines from (Bada et al., 2012), its mention should be mapped into the semantically closest concept available in the vocabulary, e.g., "B" should be mapped to *Dukes B rectal cancer*.

**Cell Type Annotation.** The cell type is the most specific superclass of all the entries in a cell group. It must first be obtained for entries with concept annotations (if they exist) using the UMLS-NCI concept hierarchy, and possibly generalised to entries without concept annotations. However, if there are no concept annotations for the cells in a cell group (54.22% of cell groups in our dataset), or the concept annotations do not have an informative superclass in UMLS-NCI (e.g., when the most specific common ancestor of concept annotations in UMLS-NCI is *Conceptual Entity*, *NCI Administrative Concept* or *NCI Thesaurus*), the annotator must retrieve the most specific concept in UMLS-NCI that best describes the content of the cells in the cell group. For example, a cell group that lists chromosomes (e.g., "1", "2", "18") is annotated with the concept *[C0008633] Chromosome*.

**Relation Annotation.** The relation annotation guidelines were developed under the assumption that the relation annotation phase would follow the cell type annotation phase. This meant that all relation hypotheses could be automatically pre-computed using the constraints from the UMLS Metathesaurus and the UMLS SN, and suggested to the annotators, who in turn could accept or reject the suggestions. Therefore, the relation annotation guidelines contain the following information: (1) definition and examples of what constitutes a valid relation between two cell groups; (2) a definition and use cases for the *isa* relation; and (3) a list of cases where no relation should be built between two cell groups.

### 3.3 Choice of Corpus Distribution Formats and Pre-processing of the Documents

When choosing formats for the articles and annotation, we considered the following criteria: (1) they must preserve information about table structure; (2) they should preferably preserve information about the structure and formatting of the original paper; and (3) they should be flexible and expressive enough to uniformly encode all the annotation types and schemas described above.

We considered various linguistic annotation formats, such as BRAT stand-off, BioC and XML/JSON stand-off. However, neither BRAT nor BioC satisfy the first two criteria, as they convert tables into plain text, and BRAT also stores documents in plain text. We therefore decided to distribute the articles in the original XML format used by PubMed Central for archiving. Such XML versions of articles use The Journal Archiving and Interchange Tag Set,[3] which preserves the content, structure and format of the articles and the tables within. The created annotations were stored in stand-off JSON format – one JSON annotation file per paper.

To construct the dataset, we downloaded the XML versions of the papers from the FTP service of PubMed Central.[4] We then automatically assigned unique IDs to all the XML tags that contained actual content (as opposed to article meta-data), such as paragraphs `<p>`, section titles `<title>`, article titles `<article-title>`, table headers `<th>` and data cells `<td>`. The templates of the JSON annotation files for each paper in the dataset were automatically generated by a script.

### 3.4 Annotation Process

The annotation process was conducted in three stages. First, cell groups were annotated, followed by the annotation of concepts and cell types for each cell group; relations between cells were annotated last.

*Cell group* annotation was carried out by the first author for the 100 tables of the dataset (43 papers) using a text editor with programming language support.

---

[3]`http://dtd.nlm.nih.gov/archiving/`
[4]`ftp://ftp.ncbi.nlm.nih.gov/pub/pmc/`

Owing to budgetary constraints, the annotation for the second and third stages was done on a subset of 83 tables in 39 papers. The annotation was performed by the first author and two specialist annotators who hold post-graduate degrees in Biology and are familiar with the subject matter. The annotators received extensive annotation guidelines for each stage, were instructed in the use of the annotation tools, and were advised to consider information from the full text of the papers during the annotation process. Thereafter, the annotators performed a double-blind annotation of a subset of five tables (two papers), to assess the level of inter-annotator agreement, and to discover any problems related to the guidelines and tools. The budgetary constraints also led us to perform a single-blind annotation for the remaining 78 tables (37 papers), which were distributed between the first author and the annotators: each participant annotated their assigned tables, which were then passed to another team member for verification. Throughout the annotation process (130 hours per person) the first author and the annotators met after every 6 to 8 hours of annotation, in order to measure the agreement between the original annotator of a table and the "reviewing" team member; annotation disagreements were resolved by consensus, and the annotation guidelines were amended where necessary, which happened rarely. At the end of each annotation stage, the master version of the annotations for the entire corpus was verified by the first author for consistency and compliance with the guidelines. Annotations were corrected where necessary.

*Concept* and *cell-type* annotations were carried out using a modified version of the BRAT annotation tool. Prior to loading the corpus into the tool, the XML files were automatically mapped into plain text (the input format of BRAT). Upon completion of the annotation, the files were mapped back into the XML/JSON distribution format.[5] We employed an in-house Web interface to integrate the NCI subset of the UMLS Metathesaurus into our BRAT installation. The annotators used BRAT to select text spans for annotation, and to query the Web interface for concepts related to these text spans. The Web interface allowed annotators to browse the lists of returned concepts, and to look up information about these concepts, such as name, STs, definition and position in the NCI concept hierarchy. If a suitable concept was found, the annotation was sent back to BRAT.

Finally, *relations between cells* were annotated using an in-house online Relation Annotation Tool, which suggested relation annotations drawn from UMLS-NCI and the UMLS Semantic Network on the basis of existing cell type annotations.

## 4 Results and Discussion

### 4.1 Corpus and Annotation Statistics

83 tables from our corpus were manually annotated with cell groups, *C*oncepts, *C*ell *T*ypes and *R*elations (denoted *CCTR-83*); 17 additional tables were annotated with cell groups only. Table 1 details the composition of the dataset. As seen in Table 1, the corpus is evenly split between two main topics, human cancer and mouse genetics (43 and 40 tables in CCTR-83 respectively), offering interesting opportunities for cross-domain training and testing. Statistics about the dimensions of the tables (average, median, minimum and maximum per table, and total counts) appear in Table 2. Table 3 shows statistics of cell-group, concept, cell-type and relation annotations, both for all annotations (left-hand side) and unique annotations (right-hand side).

As seen in Table 3, the unique concept and cell-type annotations constitute a relatively high percentage of their total counts (528 out of 3042 concepts, and 375 out of 2545 cell types based on UMLS CUIs). However, the distributions of the unique annotations are skewed. For example, the top-three most frequent cell-type annotations based on UMLS-NCI (*Count*, *Percent* and *Biologic Entity Group Quantity*) together constitute 36% of the 2545 cell types based on UMLS CUIs, while 178 cell-type annotations appear only once in the corpus. The most frequent cell-type annotation based on UMLS STs is *Quantitative Concept*, comprising 49.7% of the 2089 UMLS ST cell-type annotations; followed by the label *Organism Attribute*, which constitutes only 6% of the annotations. Such a strong bias towards *Quantitative Concept* may be explained by the predominantly quantitative nature of biomedical tables: 43.5% of the cell groups in our corpus contain numbers and numerical expressions, while 25.8% and 23.7% of the cell groups contain free text (terms and phrases) and abbreviations respectively; the remaining 7% of the

---

[5]To our knowledge, currently there is no annotation tool that natively supports table annotation. Due to our budgetary constraints, we were unable to develop such an annotation tool ourselves, and had to resort to partial suboptimal solutions.

| Source Dataset | Domain | Full dataset | | CCTR-83 | |
|---|---|---|---|---|---|
| | | Papers | Tables | Papers | Tables |
| CRAFT | Mouse genetics | 24 | 50 | 22 | 40 |
| HVP Corpus | Human colorectal cancer genetics | 9 | 28 | 8 | 24 |
| Jimeno Yepes and Verspoor (2013) | Human cancer genetics | 8 | 17 | 8 | 17 |
| Wong et al. (2009) | Human cancer genetics | 2 | 5 | 1 | 2 |

Table 1. Article and table counts and domains

| Table Element | Full dataset | | | | | CCTR-83 | | | | |
|---|---|---|---|---|---|---|---|---|---|---|
| | Total # | # per table | | | | Total # | # per table | | | |
| | | Avg. | Med. | Min. | Max. | | Avg. | Med. | Min. | Max. |
| **Cells** | 13061 | 130.61 | 78 | 12 | 1300 | 9753 | 117.51 | 64 | 12 | 1300 |
| **Rows** | 1929 | 19.29 | 13 | 3 | 100 | 1500 | 18.07 | 10 | 3 | 100 |
| **Columns** | 631 | 6.31 | 5 | 2 | 16 | 500 | 6.02 | 5 | 2 | 13 |

Table 2. Counts of table cells, rows and columns in the dataset

| Annotation Type | All annotations | | | | | Unique annotations | | | | |
|---|---|---|---|---|---|---|---|---|---|---|
| | Total # | # per table | | | | Total # | # per table | | | |
| | | Avg. | Med. | Min. | Max. | | Avg. | Med. | Min. | Max. |
| **Cell Group** | 2443 | 24.43 | 15.0 | 4 | 163 | – | – | – | – | – |
| CCTR-83 | 2134 | 25.71 | 15.0 | 4 | 163 | – | – | – | – | – |
| **Concept** | 3042 | 36.65 | 23 | 1 | 296 | 528 | 13.83 | 11 | 1 | 37 |
| **Cell Type** | | | | | | | | | | |
| UMLS CUI | 2545 | 30.66 | 20 | 4 | 148 | 375 | 11.28 | 10 | 3 | 25 |
| UMLS ST | 2089 | 25.17 | 16 | 4 | 122 | 52 | 6.75 | 6 | 2 | 15 |
| **Relation** | | | | | | | | | | |
| All labels | 1847 | 22.52 | 9 | 1 | 113 | 31 | 3.22 | 3 | 1 | 12 |
| UMLS MetaTh. | 222 | 2.71 | 0 | 0 | 32 | 4 | 0.61 | 0 | 0 | 2 |
| UMLS SN | 1625 | 19.82 | 9 | 1 | 109 | 27 | 2.61 | 2 | 1 | 10 |

Table 3. Annotation counts (including ambiguous annotations)

cell groups have mixed content (e.g., "MSI-H (n = 19)"). The breakdown for cells is 52.4% numerical, 14% text, 26.7% abbreviations and 6.9% mixed content.

Another noteworthy observation is the relatively modest corpus coverage provided by our concept and relation annotations. Concept annotations were assigned to only 30.47% of the non-empty table cells, which corresponds to 45.78% of the cell groups. This may be explained by (1) the quantitative nature of many biomedical tables combined with the difficulty of mapping numbers to concepts; and (2) the insufficient coverage of table entries by UMLS-NCI for non-numerical concepts such as specific mutations (e.g., "c.1886 A > G"), base sequences (e.g., "5'-dT20-ACTGGC...GAAAAC-3'") and patient IDs (e.g., "IC628"). With regard to relations, even after we expanded the relation annotation schema with labels from UMLS SN, we annotated only 1847 relations between the available 2134 cell groups in the CCTR-83 dataset, yielding a small set of only 31 unique labels, and a median of nine relations per 15-cell-group table (column 4 in Table 3). The distribution of the relation labels in the corpus is even more skewed than the distribution of the concept and cell-type labels, with about 75% of the relations being quite general: the *isa* relation from UMLS SN constitutes 51% of the relation labels, followed by the labels *associated_with* from UMLS SN (11.3%), and *isa* and *inverse_isa* from UMLS-NCI (each 6%). This imbalance may be attributed to the discrepancy between the information in our tables and the relations available in UMLS, which may be mitigated by employing a different annotation schema.

| Annotation | Kappa values | Kappa values (per paper) | | | |
|---|---|---|---|---|---|
| | (entire corpus) | Avg. | Med. | Min. | Max. |
| **Concept** | 0.88 | 0.90 | 0.96 | 0.13 | 1.0 |
| **Cell Type** | | | | | |
|     UMLS CUI | 0.87 | 0.82 | 0.92 | 0.15 | 1.0 |
|     UMLS ST | 0.87 | 0.86 | 0.94 | 0.23 | 1.0 |
| **Relation** | 0.82 | 0.87 | 0.91 | 0.49 | 1.0 |

Table 4. Inter-annotator agreement for concept, cell-type and relation annotation

## 4.2  Inter-Annotator Agreement

To enable the prompt resolution of problems in the annotation guidelines, and to map the progression of inter-annotator agreement (IAA) over time, per-paper IAA was measured at every conflict-resolution meeting throughout the annotation process. We used Cohen's Kappa statistic (Cohen, 1960) to evaluate IAA for all types of annotations.

Two concept annotations were deemed to match if their UMLS CUIs and text spans were equal; two UMLS CUI cell-type annotations matched if their CUIs were equal, and similarly, two UMLS ST cell-type annotations matched if their STs were equal; and two relation annotations were deemed to match if they had the same relation label, direction and arguments. It is worth noting that every cell group received at least one cell-type label; if there was more than one label (ambiguous annotation), each label was considered separately when computing IAA. In contrast, some table-entries did not have concept annotations, and similarly, some pairs of cell groups did not have relation annotations. In order to handle these cases, as well as ambiguous annotations, we added the label *No_Annotation*, so that IAA could be computed between absent and present labels.

IAAs computed over the entire corpus and per-paper IAAs appear in Table 4. For all annotation types, the average and median IAA values exceeded 0.82 and 0.91 respectively. This shows that, for most papers in our dataset, the application of our annotation guidelines yielded highly consistent annotations. However, the low minimum IAA values indicate that a few papers posed a significant challenge. This was variously due to (1) ambiguities in the annotation guidelines, which were fixed after discussing the relevant part of the guidelines;[6] (2) erroneous annotations caused by lack of clarity and ambiguity of some concepts and relations in the UMLS; and (3) absence of a concept annotation from which a cell-type annotation could be derived — these cases were less consistent across annotators than those where concept annotations were available.

## 5  Conclusion

We have offered a corpus comprising 100 tables sourced from 43 biomedical journal articles on the topic of genetics. All the tables in the corpus were manually annotated with information about homogeneous cell groups, and a subset of 83 tables was annotated with a total of more than 3000 concepts, 2000 cell types and 1800 relations, drawn from the Unified Medical Language System®. Our annotation schema was designed to accurately capture fine-grained semantic classes of table entries and the relationships between them. This annotation schema, combined with the stringent table annotation guidelines we developed, enabled a high average inter-annotator agreement of over 0.82 for all annotation types. This makes both the annotated corpus and the guidelines used to create it a valuable resource for the development and evaluation of tools for information extraction from biomedical tables. Furthermore, although our guidelines were developed for a particular biomedical corpus, they may be adapted to tables from other scientific fields, thus providing a general framework for table annotation.

## Acknowledgements

This research was supported in part by National ICT Australia (NICTA), CSIRO and the Monash University FIT Graduate Research Candidature Funding Scheme.

---

[6]After updating the guidelines at a meeting, the papers were not returned to the corresponding annotators for re-annotation, hence the IAA was not re-calculated. Only the master version of the annotations in the entire corpus was reviewed for consistency and compliance with the guidelines at the end of each annotation stage.

**References**

Michael Bada, Miriam Eckert, Donald Evans, Kristin Garcia, Krista Shipley, Dmitry Sitnikov, William A. Baumgartner, K. Bretonnel Cohen, Karin Verspoor, Judith A. Blake, and Lawrence E. Hunter. 2012. Concept annotation in the CRAFT corpus. *BMC Bioinformatics*, 13(1):161.

Sally Bamford, Todd E. Dawson, Simon A. Forbes, Jody Clements, Roger Pettett, Ahmet Dogan, Adrienne M. Flanagan, Jon Teague, P. Andrew Futreal, and Michael R. Stratton. 2004. The COSMIC (Catalogue of Somatic Mutations in Cancer) database and website. *British Journal of Cancer*, 91(2):355–358.

Michael J. Cafarella, Alon Y. Halevy, Daisy Zhe Wang, Eugene Wu, and Yang Zhang. 2008. WebTables: Exploring the power of tables on the Web. *Proceedings of the VLDB Endowment*, 1(1):538–549.

David Campos, Sérgio Matos, and José Luís Oliveira. 2013. A modular framework for biomedical concept recognition. *BMC Bioinformatics*, 14:281.

Jacob Cohen. 1960. A coefficient of agreement for nominal scales. *Educational and Psychological Measurement*, 20(1):37–46.

Rezarta Islamaj Doğan, Robert Leaman, and Zhiyong Lu. 2014. NCBI disease corpus: A resource for disease name recognition and concept normalization. *Journal of Biomedical Informatics*, 47:1–10.

Gaëlle Hignette, Patrice Buche, Juliette Dibie-Barthélemy, and Ollivier Haemmerlé. 2009. Fuzzy annotation of web data tables driven by a domain ontology. In *Proceedings of the 6th European Semantic Web Conference (ESWC 2009)*, pages 638–653, Heraklion, Crete, Greece. Springer-Verlag.

Jane Hoover-Plow, Aleksey Shchurin, Erika Hart, Jingfeng Sha, Annie E. Hill, Jonathan B. Singer, and Joseph H. Nadeau. 2006. Genetic background determines response to hemostasis and thrombosis. *BMC Hematology*, 6(1):1.

Matthew Francis Hurst. 2000. *The interpretation of tables in texts*. Ph.D. thesis, University of Edinburgh.

Dietmar Jannach, Kostyantyn Shchekotykhin, and Gerhard Friedrich. 2009. Automated ontology instantiation from tabular web sources – the AllRight system. *Web Semantics: Science, Services and Agents on the World Wide Web*, 7(3):136–153.

Antonio Jimeno Yepes and Karin Verspoor. 2013. Towards automatic large-scale curation of genomic variation: improving coverage based on supplementary material. In *Proceedings of BioLINK SIG 2013: Roles for text mining in biomedical knowledge discovery and translational medicine*, pages 39–43, Berlin, Germany.

Jin-Dong Kim, Tomoko Ohta, Yuka Tateisi, and Junichi Tsujii. 2003. GENIA corpus – a semantically annotated corpus for bio-textmining. *Bioinformatics*, 19(suppl 1):i180–i182.

Kyubum Lee, Sunwon Lee, Sungjoon Park, Sunkyu Kim, Suhkyung Kim, Kwanghun Choi, Aik Choon Tan, and Jaewoo Kang. 2016. BRONCO: Biomedical entity relation oncology corpus for extracting gene-variant-disease-drug relations. *Database: The Journal of Biological Databases and Curation*, 2016.

Girija Limaye, Sunita Sarawagi, and Soumen Chakrabarti. 2010. Annotating and searching web tables using entities, types and relationships. *Proceedings of the VLDB Endowment*, 3(1-2):1338–1347.

Varish Mulwad, Tim Finin, and Anupam Joshi. 2013. Semantic message passing for generating linked data from tables. In *Proceedings of the 12th International Semantic Web Conference (ISWC 2013)*, pages 363–378, Sydney, Australia. Springer-Verlag.

John-Paul Plazzer, Rolf H. Sijmons, Michael O. Woods, Päivi T. Peltomäki, Bryony A. Thompson, Johan T. Den Dunnen, and Finlay Macrae. 2013. The InSiGHT database: utilizing 100 years of insights into Lynch Syndrome. *Familial Cancer*, 12(2):175–180.

Gianluca Quercini and Chantal Reynaud. 2013. Entity discovery and annotation in tables. In *Proceedings of the 16th International Conference on Extending Database Technology (EDBT '13)*, pages 693–704, Genoa, Italy. ACM.

Dietrich Rebholz-Schuhmann, Antonio José Jimeno Yepes, Erik M. van Mulligen, Ning Kang, Jan A. Kors, David Milward, Peter T. Corbett, Ekaterina Buyko, Katrin Tomanek, Elena Beisswanger, and Udo Hahn. 2010. The CALBC silver standard corpus for biomedical named entities – a study in harmonizing the contributions from four independent named entity taggers. In *Proceedings of the 7th International Conference on Language Resources and Evaluation (LREC '10)*, pages 568–573, Valletta, Malta. European Language Resources Association.

Barbara Rosario and Marti A. Hearst. 2004. Classifying semantic relations in bioscience texts. In *Proceedings of the 42nd Annual Meeting of the Association for Computational Linguistics (ACL '04)*, pages 430–437, Barcelona, Spain. Association for Computational Linguistics.

Tatyana Shmanina, Lawrence Cavedon, and Ingrid Zukerman. 2014. Challenges in information extraction from tables in biomedical research publications: A dataset analysis. In *Proceedings of the Australasian Language Technology Association Workshop 2014 (ALTA 2014)*, pages 118–122, Melbourne, Australia. Association for Computational Linguistics.

Nicholas Sioutos, Sherri de Coronado, Margaret W. Haber, Frank W. Hartel, Wen-Ling Shaiu, and Lawrence W. Wright. 2007. NCI Thesaurus: A semantic model integrating cancer-related clinical and molecular information. *Journal of Biomedical Informatics*, 40(1):30–43.

Mark Van Assem, Hajo Rijgersberg, Mari Wigham, and Jan Top. 2010. Converting and annotating quantitative data tables. In *Proceedings of the 9th International Semantic Web Conference (ISWC 2010)*, pages 16–31, Shanghai, China. Springer-Verlag.

Petros Venetis, Alon Y. Halevy, Jayant Madhavan, Marius Paşca, Warren Shen, Fei Wu, Gengxin Miao, and Chung Wu. 2011. Recovering semantics of tables on the Web. *Proceedings of the VLDB Endowment*, 4(9):528–538.

Karin Verspoor, Antonio Jimeno Yepes, Lawrence Cavedon, Tara McIntosh, Asha Herten-Crabb, Zoë Thomas, and John-Paul Plazzer. 2013. Annotating the biomedical literature for the human variome. *Database: The Journal of Biological Databases and Curation*, 2013.

Jingjing Wang, Haixun Wang, Zhongyuan Wang, and Kenny Q. Zhu. 2012. Understanding tables on the Web. In *Proceedings of the 31st ER International Conference on Conceptual Modeling (ER 2012)*, pages 141–155, Florence, Italy. Springer-Verlag.

Wern Wong, David Martinez, and Lawrence Cavedon. 2009. Extraction of named entities from tables in gene mutation literature. In *Proceedings of the BioNLP 2009 Workshop (BioNLP '09)*, pages 46–54, Boulder, Colorado, USA. Association for Computational Linguistics.

Mohamed Yakout, Kris Ganjam, Kaushik Chakrabarti, and Surajit Chaudhuri. 2012. InfoGather: Entity augmentation and attribute discovery by holistic matching with web tables. In *Proceedings of the 2012 ACM SIGMOD International Conference on Management of Data (SIGMOD '12)*, pages 97–108, Scottsdale, Arizona, USA. ACM.

Xiaoxin Yin, Wenzhao Tan, and Chao Liu. 2011. FACTO: a fact lookup engine based on web tables. In *Proceedings of the 20th International Conference on World Wide Web (WWW '11)*, pages 507–516, Hyderabad, India. ACM.

Mohamed Amir Yosef, Johannes Hoffart, Ilaria Bordino, Marc Spaniol, and Gerhard Weikum. 2011. AIDA: An online tool for accurate disambiguation of named entities in text and tables. *Proceedings of the VLDB Endowment*, 4(12):1450–1453.

# Supervised classification of end-of-lines in clinical text with no manual annotation


**Pierre Zweigenbaum**
LIMSI, CNRS,
Université Paris-Saclay
91405 Orsay, France
pz@limsi.fr

**Cyril Grouin**
LIMSI, CNRS,
Université Paris-Saclay
91405 Orsay, France
grouin@limsi.fr

**Thomas Lavergne**
LIMSI, CNRS, Univ. Paris-Sud,
Université Paris-Saclay
91405 Orsay, France
lavergne@limsi.fr


## Abstract


In some plain text documents, end-of-line marks may or may not mark the boundary of a text unit (e.g., of a paragraph). This vexing problem is likely to impact subsequent natural language processing components, but is seldom addressed in the literature. We propose a method which uses no manual annotation to classify whether end-of-lines must actually be seen as simple spaces (soft line breaks) or as true text unit boundaries. This method, which includes self-training and co-training steps based on token and line length features, achieves 0.943 F-measure on a corpus of short e-books with controlled format, F=0.904 on a random sample of 24 clinical texts with soft line breaks, and F=0.898 on a larger set of mixed clinical texts which may or may not contain soft line breaks, a fairly high value for a method with no manual annotation.


## 1 Introduction

Text segmentation is a low-level task which contributes to the higher-level information extraction tasks performed by natural language processing; for instance, Smith (2011, p. 5) states that *"If we build a language model on poorly segmented text, for instance, its predictive performance will suffer."* Specifically, splitting a text into sentences, despite its looking like a largely solved problem, continues to raise nagging issues for some ill-formatted texts such as clinical texts (Miller et al., 2015). Most methods and software performing higher-level tasks (e.g., cTAKES (Savova et al., 2010) and others), such as part-of-speech tagging, syntactic parsing, entity and relation extraction, depend on low-level processes such as sentence segmentation. This paper focuses on a little-addressed, basic component in the NLP pipeline, which impacts sentence splitting and hence subsequent processes. This component may be seen as the determination of *paragraph boundaries*, or the *classification of end-of-lines*.

The problem can be described as follows. In some plain text documents, such as e-mail messages, text fields in databases, or PDF documents converted into text, the line break or end-of-line mark may or not play the role of a boundary marker for a text unit (a title, a paragraph, etc.) and hence may or not mark a sentence boundary. In some text documents the end-of-line mark is always a paragraph (or title) boundary, and no problem occurs: subsequent processes such as sentence splitting can be run within each paragraph. But in some text documents, an end-of-line mark may occur in the midst of a paragraph, typically to "wrap" paragraphs that exceed some set length: depending on the origin of the text and on input and formatting conditions, this may have been caused by an automatic process ('hard' line wrapping in some text editors) or by manual intervention of the typist. Often enough these originating conditions are not precisely known at the time these documents are submitted to natural language processing. Preprocessing must then address this situation and include some solution to the classification of end-of-line marks (henceforth noted <EOL>), i.e., to determine whether an <EOL> must be considered as an actual text unit boundary (henceforth <TUB>) or should be considered as standing for a simple space (<SP>), meaning that this line has incurred "paragraph wrapping" and should be considered together with (e.g., pasted to) the next line to form a larger text unit.






*Proceedings of the Fifth Workshop on Building and Evaluating Resources for Biomedical Text Mining (BioTxtM 2016),*
pages 80–88, Osaka, Japan, December 12th 2016.

This situation is mentioned by some authors, e.g., by Miller et al. (2015) for the MIMIC II clinical texts, or by Zweigenbaum & Grouin (2014) for the i2b2/UTHealth 2014 NLP challenge documents (Stubbs et al., 2015). It is probably present in a much larger set of text collections, hence is likely to create some problems for many systems and teams working with these documents. While the impact of these problems still needs to be assessed precisely (see, e.g., (Zweigenbaum and Grouin, 2014) for limited examples), and may depend on the type of processing that follows, the number of situations where they are likely to occur warrants an investigation into general methods to address the task of classifying <EOL> marks.

End-of-line classification is therefore the topic of the present paper. We address it as a problem in itself, independently of its impact on subsequent tasks, and thus perform intrinsic evaluations of its performance; extrinsic evaluations, for instance through its impact on sentence splitting accuracy, are left for future work. Because <EOL> classification, although often needed, is only a small piece of preprocessing in a larger natural language processing pipeline whose adaptation to a given clinical task generally already requires some human annotation effort, a supervised method requiring more human annotation is not desirable. We therefore endeavored to investigate methods which require no human annotation to perform this task.

## 2   Related work

The problem of <EOL> classification seems to be little explored in natural language processing (NLP), and the section that Smith (2011) dedicates to segmentation does not mention it. Some NLP research (Sporleder and Lapata, 2006; Filippova and Strube, 2006) has addressed paragraph segmentation from a quite different perspective: given a text split into sentences, determine paragraph boundaries. However, they started from texts where sentence boundaries were given, and the input texts were assumed to be "clean" from the point of view of <EOL> marks (i.e., either sentence boundaries are deterministically marked by <EOL>s or by XML markup). A few papers on clinical NLP have recently addressed it and proposed methods based upon heuristics and knowledge about the usual format of the texts (Zweigenbaum and Grouin, 2014) or supervised machine learning (Miller et al., 2015).

Some document analysis research considers the notion of paragraph when converting a document into text by optical character recognition (Radakovic et al., 2013) or to reformat a text obtained from a PDF file (Fang et al., 2011). To perform these tasks they need to decide whether or not an <EOL> marks a paragraph boundary: this problem is similar to the one we address here. Radakovic et al. (2013) check whether a line starts with certain symbols (e.g., bullet points) or character case (uppercase or lowercase letters), ends with a number, together with other clues related to the number of words in the line, to its left and right indent size, to character font size, line coordinates in the page, distance between lines, and the presence of images. Fang et al. (2011) use information about the vertical and horizontal positioning of characters in the page, which are not available in plain text, together with paragraph indenting information. In contrast to (Radakovic et al., 2013), they do not use clues obtained from text content. To determine the reading order in a set of text objects (lines, paragraphs, etc.), Aiello et al. (2002) combine information on the spatial positioning of these objects and on the probability that the part-of-speech tag of the first word in the next object follows those of the last two words in the current object, according to a language model. This is the only reference we found in document analysis where a language model is used to help decide whether or not two lines must follow each other (which would mean, for us, be merged within the same paragraph).

We address a situation where we are given plain text but no information on the original layout of the page, such as spatial positioning of lines or characters, font size, actual left or right indent size of each line in its displayed form, which play a key role in document analysis methods. Text content, in contrast, is readily exploitable: language models based on the distribution of word features at the beginning and end of lines, as well as the distribution of line lengths in documents, can be used as cues.

Additionally, we aim to find methods which involve no human annotation. We shall see in the following section that a method consists in framing the situation as a supervised learning problem where each whitespace in a text (including spaces and <EOL>s) must be classified as a simple space (<SP>) or a text

| Corpus (Gutenberg identifier) | Language | Documents | Paragraphs | Lines (wn) | Words |
|---|---|---|---|---|---|
| Around the World in 80 Days (103) | en | 37 | 1679 | 6106 | 62,752 |
| Around the World in 80 Days (2154) | en | 37 | 1734 | 6887 | 65,510 |
| Le Tour du Monde en 80 Jours (800) | fr | 37 | 2053 | 6962 | 66,878 |
| De la terre à la lune (38674) | fr | 28 | 1449 | 5609 | 53,724 |
| Da terra à lua (28341) | pt | 28 | 1503 | 5869 | 57,487 |
| Reis naar de Maan in 28 dagen… (27309) | du | 47 | 1905 | 6915 | 65,671 |
| Around the World in 80 Days (103+2154) | en | 74 | 3413 | 12,993 | 128,262 |
| Le Tour (800) + De la terre (38674) | fr | 65 | 3502 | 12,571 | 120,602 |
| i2b2/UTHealth 2014 training corpus | en | 790 | — | 73,590 | 488,904 |
| test subset | en | 64 | 3554 | 5619 | 38,167 |

Table 1: Corpora. Lines are measured on the *wn* version of the documents, paragraphs on the *ln* version.

unit boundary (<TUB>), where part of the training examples (spaces) are positive and the rest (<EOL>) are unannotated. Nigam et al. (2000) show how text classification obtained by a Naive Bayes classifier can be improved by exploiting unannotated data on top of annotated data. This is how they train a classifier on texts whose class is known, then use it in a self-training fashion to compute the probabilities of all classes for each unannotated text. The additional information thus obtained allows them to re-train the classifier then to iterate until convergence, according to the expectation-maximization algorithm (EM). The method we propose below to train an <EOL> classifier is related to this principle, but does not need an initial human annotation. Elkan and Noto (2008) propose a non-iterative method for this purpose, but it assumes that the annotated examples are drawn randomly from the positive examples, which is not the case in our situation. Yet another path would consist in considering the <EOL> annotations as ambiguous (both <SP> and <TUB>) and in applying the methods of (Wisniewski et al., 2014). However, this would create a systematic dependency between these two classes in these annotations, a situation in which learning is not guaranteed (Bordes et al., 2010).

## 3 Material and methods

### 3.1 Corpora

We target here clinical texts with a complex mixture of formats. However, we also test our methods on more controlled corpora which we have in several formats. The controlled-format corpora are made of six plain text e-books by Jules Verne in four languages from the Gutenberg project (http://www.gutenberg.net), which we split into chapters. Each of their paragraphs is split into multiple lines (wrapped) if it exceeds a given threshold, and is bounded by blank lines. We consider each of these e-books, and the combination of the two English e-books and that of the two French e-books (see Table 1).

We produced versions of the e-books in which two properties of the paragraphs were set. First, a paragraph can be wrapped (*w*: broken into separate lines, the original format in this case) or typeset as one long line (*l*). Second, paragraph boundaries can be marked with a blank line (*b*) or not (*n*). This results in four combined formats. Among these, *wn* is the most difficult format to handle: it is the only one which has no simple paragraph delimitation. This is the one our system is meant to address. In *wb*, apart from obvious blank line separators, all end of lines should be classified as <SP>s: it is used to test whether our system produces false negatives. Conversely, *ln* has no obvious separators but no wrapped line at all: it is used to test whether our system produces false positives. Finally, *lb* has no wrapped line at all, and blank line separators everywhere, so our system handles it perfectly without classification.

The clinical text corpus is the i2b2/UTHealth 2014 NLP challenge (Stubbs et al., 2015) training corpus, which contains 790 records. These records keep the layout of a printed document and can include fixed-width columns, blank lines between text lines to reproduce double spacing, approximate positioning of elements on the page (tabulation, multiple spaces), table column separators represented by a special character (circumflex accent, pipe), etc. Some of the files have wrapped paragraphs, other do not.

We first preprocessed these texts to handle what is probably double line-spacing documents. To focus on the paragraph wrapping problem, we handled double line-spacing deterministically by removing every other blank line in texts whose ratio of contiguous pairs of non-blank lines over the number of blank lines is below an empirical ceiling of 10%.

For evaluation during the development and test of our present <EOL> classifier, we manually annotated <EOL>s in a randomly sampled 64-document subset of the i2b2 training corpus. After initial annotation of another, smaller sample by two annotators and observation of a near-perfect inter-annotator agreement, we decided that this subset could be annotated by only one annotator. Note that since our method uses no manual annotation, the human annotations in these documents were not used to train the system, only to evaluate it. Specifically, using a plain text editor (emacs), we marked each <EOL> with a code as follows:

- 0 (no paragraph wrapping) means this <EOL> is a <TUB>.

- 1 (paragraph wrapping) means this <EOL> should be considered as an <SP>.

- 2 means there is no ambiguity in the present <EOL> (this is further explained below): it must be considered as a <TUB>, but no problem needs to be solved nor evaluated here.

### 3.2   Task modeling

We decompose the overall task of end-of-line classification into two parts:

1. Determine whether a document is subject to paragraph folding, i.e., whether it is liable to contain at least one <EOL> which should be categorized as an <SP>.

2. In a document which is subject to paragraph folding, classify <EOL>s as <SP> or <TUB>.

We focus here on the second part of the task, <EOL> classification proper, assuming that the first (easier) part, document classification, is solved, for instance in a supervised way. We also present a direction to address document classification with no human annotation, which we integrate into our general method.

### 3.3   Determination of documents with folded paragraphs

A text in which paragraphs are folded is likely to have a number of lines with similar lengths, which should thus be close to the mean line length in this text. Conversely, a text in which some lines are much longer than most other lines is probably not subject to paragraph folding, otherwise these longer lines would have been folded. The distribution of line lengths in a text, compared to their mean length, should therefore be a useful clue to determine whether a text is likely to have incurred paragraph folding.

Zweigenbaum and Grouin (2014) used the *coefficient of variation* of line lengths in a document (more detail is given below in the description of individual features). They set a threshold in a supervised way, over which a document was considered not to incur paragraph folding: documents with a larger variation in line lengths have a higher coefficient of variation, whereas documents with many lines of similar length have a smaller coefficient of variation. We use the same feature in the present work, but as a discretized value and with no manual annotation.

### 3.4   Blank line handling

We first solve an easy case which requires no further classification effort: a blank line, i.e., a line which is empty or only contains whitespace characters, such as whitespace and tabulations, is always considered as marking a paragraph boundary (as explained above, double line-spacing is removed if present). There is therefore no need for the classifier to learn to detect these lines: the <EOL> which ends a blank line, as well as the <EOL> of the preceding line, are both unambiguous. They are tagged with the "2" class in the training corpus and are excluded from the evaluation. These two <EOL>s remain useful however to train the classifier, because they participate in the estimation of the probabilities of occurrence of the preceding or following tokens given a <TUB> class.

### 3.5 End-of-line classification by self-training on noisy data and and co-training

The input to this task is made of tokenized texts, where punctuation has been separated from words. These texts contain whitespace spans which can play the role of simple spaces (<SP>) or text unit boundaries (<TUB>). We define the task as deciding, for each <EOL>, whether it should be considered as an <SP> (i.e., this is an <EOL> which is the result of paragraph folding and should thus be converted to a space) or a <TUB> (i.e., this is a true text unit boundary).

We address this task through a first subtask which consists in learning, *for each whitespace span* in a document (whether space or <EOL>), whether it should be classified as an <SP> or an <TUB>. The peculiarities of the training data and processes in this learning task are the following:

- <EOL>s are ambiguous: we do not know at this stage whether they are true <TUB>'s or actually <SP>'s. This results in a partial annotation of the corpus: spaces are unambiguously tagged as <SP>, but we do not know the actual tags for <EOL>s. We convert this situation into a *noisy annotation* by tagging every <EOL> with <TUB>: these tags are sometimes correct and sometimes incorrect.

- Because only <EOL> marks are ambiguous in our overall task, we are only interested in applying our classifier to them, not to spaces. Training is performed on these noisy annotations, using token-based features as described below, and learns a model $M_A$ for tagging <EOL>s.

- Model $M_A$ is applied to the <EOL>s in the training corpus itself, resulting in (noisily) disambiguated <EOL> tags (self-training).

- A second model $M_B$ is learned on these new annotations, using different features (co-training). $M_B$ is applied to the <EOL>s in the training corpus, resulting in modified <EOL> tags.

- Alternatively, a combined model $M_{A.B}$ is built and applied to the <EOL>s in the training corpus.

We thus obtain, by exploiting only the naturally available information, three models and their associated <EOL> annotations. The process could be iterated, but we leave that for future work.

### 3.6 Features

We characterize a whitespace position $s_i$ (space or <EOL>) between two words in a document $d$ with the four discrete features below:

$A_1$ Left token: as in (Radakovic et al., 2013), we assume some tokens or punctuations are more often found at the end of a paragraph, while others are less often found there.

$A_2$ Right token ; similarly, we expect that some tokens or punctuations are often found at the start of a paragraph, or on the contrary are seldom found in this position.

$A_3$ Typographic form of the left token: all uppercase, capitalized, is a number, only contains punctuation (possibly differentiating between strong punctuations, i.e., period, exclamation mark and question mark, and the other punctuations), is a number followed by at least one punctuation and possibly preceded by one punctuation (typical form of bullet points). We assume that some typographic patterns, such as uppercase, are more frequent at the beginning of paragraphs, whereas others, such as strong punctuations, are more frequent at the end of paragraphs.

$A_4$ Typographic form of the right token.

These four features define a bigram language model, where a bigram is made of a whitespace position and an adjacent token or typographic form. Extension to longer n-grams is left for future work.

Additionally, we characterize a position $s_i$ at the end of a line by the following two features:

$B_1$ $l$, length in characters of the line that ends with the space $s_i$, normalized (centered and reduced) as $l_{norm}$ in document $d$ (Equation 1a). We assume that a short line (or a very long line) is unlikely to need to be "pasted back" to the following line.

$$(a) \quad l_{norm} = \frac{l - \mu_d}{\sigma_d}, \qquad \mu_d = \frac{1}{N}\sum_{l \in d} l, \qquad \sigma_d = \sqrt{E[(l - \mu_d)^2]}, \qquad (b) \quad cv_d = \frac{\sigma_d}{\mu_d} \quad (1)$$

$\mu_d$ is the mean line length in document $d$ and $\sigma_d$ is the standard deviation of line lengths in $d$.

$B_2$ $cv_d$, coefficient of variation of line length $l$ in document $d$ (Equation 1b). This feature is common to all positions in document $d$. We assume that a document whose paragraphs are folded is likely to contain a number of lines of comparable lengths (probably close to the width of the input screen or original printable page). This should result in a low standard deviation of line length compared to the mean line length. The *coefficient of variation cv* is defined as their ratio (Equation 1b).

The latter two features $l$ and $cv_d$ have numeric values, we discretize them into ten bins between their minimal and maximal values as observed in the training corpus. If a test document has out-of-range values, they are discretized into the closest bin.

### 3.7 Naive Bayes classification

We use a very simple classifier, the *Naive Bayes* classifier, which is well-known for its robustness and speed. The probability of having a given class $c_j \in C$ (here, $C = \{$<TUB>, <SP>$\}$) for a certain whitespace $s$, characterized by features (e.g., $a_i \in A$), is (2a):

$$(a) \quad P(c_j|s) = \frac{P(c_j)P(s|c_j)}{P(s)}, \qquad (b) \quad P(s|c_j) = \prod_{i=1}^{|A|} P(a_i|c_j) \qquad (2)$$

The Naive Bayes classifier hypothesizes the independence of the observed features for a given class. This leads to (2b). At inference time, the selected class is the one with the maximum a posteriori probability; since in (2a) $P(s)$ does not vary with the class $c_j$, we obtain (3a):

$$(a) \quad \arg\max_j P(c_j|s_i) = \arg\max_j P(c_j)\prod_{i=1}^{|A|} P(a_i|c_j), \qquad (b) \quad \hat{P}(a_i|c_j) \approx \frac{occ(a_i, c_j) + 1}{occ(c_j) + |C|} \qquad (3)$$

Concretely, we compute the likelihood ratio of $c_{\text{<SP>}}$ over $c_{\text{<TUB>}}$, i.e., $\frac{P(c_{\text{<SP>}}|s_i)}{P(c_{\text{<TUB>}}|s_i)}$ and decide for an <SP> (tag 1) if greater than zero, <TUB> (tag 0) otherwise.

Equation (3a) relies on an estimation of $P(c_j)$ and $P(a_i)$ in a training corpus. This is classically performed according to a maximum likelihood principle: $\hat{P}(c_j) \approx \frac{occ(c_j)}{occ(s)}$ and $\hat{P}(a_i|c_j) \approx \frac{occ(a_i, c_j)}{occ(c_j)}$ where $occ(s)$ is the total number of spaces in the corpus, $occ(c_j)$ is the number of spaces with class $c_j$, and $occ(a_i, c_j)$ is the number of spaces with feature $a_i$ and class $c_j$. A commonly encountered problem is the presence of test examples whose feature values have no occurrence in the training corpus. We address it with a widespread method, Laplace smoothing (Manning et al., 2008) (see Equation 3b above).

Training for model $M_A$ is performed on all spaces with language model features $A_1 \ldots A_4$, whereas training for model $M_B$ is performed only on <EOL>s with length features $B_1 B_2$. A study of the scores of the two models on the development corpus shows that their distributions have similar dynamics, thus that a combination of the two can be considered. We create this combination by multiplying the likelihood ratios of $M_A$ and $M_B$ for the same space, hence its name $M_{A \cdot B}$. A more sophisticated combination might improve performance, but would require an annotated corpus to optimize its parameters.

## 4 Results

Since we use no human annotation, we trained our model on the whole set of texts in each sub-corpus and applied it to the same sub-corpus: each e-book of Section 3.1, and the 790 texts in the i2b2 corpus. For the i2b2 corpus, we only have gold annotations for a 64-text subset, on which performance is measured.

| Corpus | Model | Acc | P | R | F | Acc | P | R | F |
|---|---|---|---|---|---|---|---|---|---|
| | | six individual e-books | | | | two merged e-books | | | |
| e-books: wb | wrap-none | 0 | — | 0 | 0 | 0 | — | 0 | 0 |
| e-books: wb | wrap-all | 1.000 | 1.000 | 1.000 | 1.000 | 1.000 | 1.000 | 1.000 | 1.000 |
| e-books: wb | $M_A$ | 0.733 | 1.000 | 0.733 | 0.846 | 0.812 | 1.000 | 0.812 | 0.895 |
| e-books: wb | $M_B$ | **0.949** | 1.000 | **0.949** | **0.973** | 0.992 | 1.000 | **0.992** | **0.996** |
| e-books: wb | $M_{A \cdot B}$ | 0.833 | 1.000 | 0.833 | 0.908 | 0.902 | 1.000 | 0.902 | 0.948 |
| e-books: ln | wrap-none | 1.000 | — | — | — | 1.000 | — | — | — |
| e-books: ln | wrap-all | 0 | 0 | — | 0 | 0 | 0 | — | 0 |
| e-books: ln | $M_A$ | 0.996 | — | — | — | 0.995 | — | — | — |
| e-books: ln | $M_B$ | 0.833 | — | — | — | **1.000** | — | — | — |
| e-books: ln | $M_{A \cdot B}$ | **0.999** | — | — | — | 0.999 | — | — | — |
| e-books: wn | wrap-none | 0.262 | — | 0 | 0 | 0.264 | — | 0 | 0 |
| e-books: wn | wrap-all | 0.738 | 0.738 | 1.000 | 0.849 | 0.736 | 0.736 | 1.000 | 0.848 |
| e-books: wn | $M_A$ | 0.802 | **0.998** | 0.733 | 0.845 | 0.861 | **0.998** | 0.812 | 0.895 |
| e-books: wn | $M_B$ | **0.917** | 0.939 | **0.949** | **0.943** | **0.932** | 0.922 | **0.992** | **0.955** |
| e-books: wn | $M_{A \cdot B}$ | 0.875 | **0.998** | 0.833 | 0.907 | 0.924 | 0.994 | 0.902 | 0.945 |
| | | +wrap | | | | all | | | |
| i2b2 | wrap-none | 0.459 | — | 0.000 | 0.000 | 0.804 | — | 0.000 | 0.000 |
| i2b2 | wrap-all | 0.541 | 0.541 | 1.000 | 0.702 | 0.196 | 0.196 | 1.000 | 0.328 |
| i2b2 | $M_A$ | 0.900 | 0.890 | 0.930 | 0.910 | 0.846 | 0.565 | 0.930 | 0.703 |
| i2b2 | $M_B$ | 0.897 | 0.905 | 0.904 | 0.904 | **0.960** | **0.893** | 0.904 | **0.898** |
| i2b2 | $M_{A \cdot B}$ | **0.919** | **0.916** | **0.937** | **0.926** | 0.903 | 0.685 | **0.937** | 0.791 |

Table 2: Experiments on multi-format e-books and on the i2b2 evaluation corpus (+wrap = only files with paragraph wrapping); Acc = accuracy, P = precision, R = recall, F = F-measure. Note that *e-books: ln* has no wrapped paragraph hence no positive space hence no true positive, therefore it has P=R=F=0.

We evaluate results by measuring the classical accuracy, precision, recall, and F-measure, for a task whose goal is to detect whether an <EOL> should be considered a <SP>. A true positive (TP) corresponds to an <EOL> which is correctly classified as a <SP>. A false positive (FP) is an <EOL> incorrectly classified as an <SP>. A false negative (FN) occurs when an <EOL> is incorrectly classified as a <TUB>. Accuracy, precision, recall, and F-measure stem from these definitions.

We provide two simple baselines: wrap-none considers that every <EOL> is a <TUB>, and wrap-all considers that every <EOL> is a <SP>. They enable us to show the 'lift' brought by $M_A$.

Table 2 shows evaluation results for two series of experiments: ($i$) on the e-book corpus, with wrapped paragraphs separated by blank lines (*wb*), long-line paragraphs with no separating blank line (*ln*), and wrapped paragraphs with no separating blank line (*wn*); and ($ii$) on the i2b2 corpus, on paragraph-folded documents (+wrap) then on the whole Test corpus (all).

## 5  Discussion

The first baseline method *wrap-none* always has null recall and F-measure by definition. It obtains perfect accuracy when no paragraph is wrapped (*ebooks:ln*), and sets a baseline with a fair accuracy on the mixed-type *i2b2:all* corpus. The second baseline method *wrap-all* has perfect or null performance on the artificial *ebooks:wb* and *ebooks:ln* corpora respectively. It sets a useful baseline for the highly wrapped *ebooks:wn* and *i2b2+wrap* corpora. However, for the more difficult, mixed *i2b2 all* corpus, it performs poorly. $M_A$ and subsequent models outperform these baselines in F-measure and accuracy on the highly wrapped *ebooks:wn* corpus (except the F-measure of $M_A$ on *ebooks:wn*, which is only on par with that of *wrap-all* and on the i2b2 corpus).

On e-books, the best recall in condition *wb* is obtained by model $M_B$. Its accuracy looks lower in condition *ln*, but this is due to its complete failure on one document (Le tour du monde (800)), which warrants further exploration; on all other documents it creates no false positive and hence outperforms the other two models. In condition *wn*, the more difficult situation, $M_B$ obtains the best recall and F-measure, whereas $M_A$ and $M_{A \cdot B}$ obtain near-perfect precision. The merged documents (103+2154 and 800+3874, right pane of Table 2) obtain better results than each individual document they contain. Globally, recall increases by 5–8pt while precision remains constant or slightly decreased, resulting in a 1–5pt increase in F-measure. Specifically, on the merged French documents, the failure of $M_B$ which occurred on one of them disappears. Training size is thus an important factor. Not shown for reasons of space: apart from one exception mentioned above, performance was similar on the four languages (English, French, Portuguese and Dutch). The format of the texts and the consistency of their vocabulary are probably more important than the language, and this set of languages does not exhibit a wide morphological variation. In summary, in each of these tests on one type of text with very regular wrapped text, the length-based model $M_B$, trained on the output of $M_A$, performs best.

Our evaluation on i2b2 documents is unfortunately not directly comparable to the F=0.965 of (Zweigenbaum and Grouin, 2014): they used a different subset of the i2b2 corpus which is not public. Besides, they implemented an extensive set of heuristics, whereas our method relies on no heuristic and aims at being more generic. Miller et al. (2015) used human annotations on a different corpus, and their results are thus not directly comparable to the present ones either.

The documents of the i2b2 evaluation corpus which incur paragraph wrapping display a more complex distribution of *w/l* and *b/n* formats, even within one document. The three models perform well, with a small advantage to the combined model $M_{A.B}$. On the full i2b2 evaluation corpus, the length-based model $M_B$ brings a substantive improvement over $M_A$ and fares the best in terms of precision and F-measure. We hypothesize that this is due to the ability of its coefficient of variation feature to detect texts that are not likely to incur paragraph wrapping, thereby integrating a partial solution to this document classification subtask. In this setting the combined model $M_{A.B}$ strongly improves the precision and F-measure of model $M_A$: the detection of non-paragraph-wrapping texts by the length-based features removes a large number of false positives and slightly improves recall at the same time.

In summary, if one wants robustness on both types of texts with a balanced precision, recall, and F-measure, the length-based model $M_B$, trained with annotations obtained from the token-based model $M_A$, is the most stable choice, at or above 0.90 for all these measures. Its features help the language-model model $M_A$ reach a better precision and F-measure for the documents which are not subject to paragraph wrapping. However, if one looks for high-recall detection of wrapped paragraphs, model $M_A$ has a higher recall of 0.93, and its combination with $M_B$ was the most successful on wrapped texts. Therefore, depending on the needs of the natural language processing task that will be run on these texts, the choice of the model allows to favor precision ($M_B$) or recall ($M_A$ or, better, $M_{A.B}$). Besides, models ($M_A$ or, better, $M_{A.B}$) show their true performance on texts with paragraph wrapping. Finally, the combination $M_{A.B}$, albeit efficient on i2b2 wrapped texts, was not sufficient to block false positives on texts with no wrapped paragraphs. Other strategies might be more successful, such as using $M_B$ in a first step as a filter to detect and exclude non-wrapped documents, hence restricting the application of $M_A$ or $M_{AB}$ to an automatically detected +wrap subset.

A most interesting perspective is the study of the interaction of <EOL> classification with sentence segmentation. On the one hand, as suggested by one of the reviewers, sentence segmentation might be used as a baseline for <EOL> classification, all the more in texts where paragraphs typically end with a period. On the other hand, the study of the impact of <EOL> classification on sentence segmentation is one of the motivations for the present work, and constitutes our next step. As suggested by another reviewer, section title detection (Tepper et al., 2012) can also help paragraph segmentation. As a matter of fact, it was part of the heuristics used in (Zweigenbaum and Grouin, 2014), where it helped to avoid pasting a title (possibly with no final period) to the next line.


## 6   Conclusion

We presented a method which uses self-training and co-training to classify <EOL>s with no human annotation, based on available token and line length features. It achieves high <EOL> classification F-measures on i2b2 clinical texts which incur paragraph folding, and can also detect texts which are not subject to this phenomenon.

In future work, we plan to test $M_B$ as a filter as outlined above. We will also explore other features such as POS tags and n-grams with $n > 1$, more powerful classifiers such as logistic regression and SVM, and perform extrinsic evaluations such as the impact on sentence segmentation.

**Acknowledgements**

This work was partially funded by BPI under FUI-15 grant SONAR and by the European Union's Horizon 2020 Marie Sklodowska Curie Innovative Training Networks—European Joint doctorate (ITN-EJD) under grant agreement No:676207, Methods in Research on Research (MiRoR).

**References**

Marco Aiello, Christof Monz, Leon Todoran, and Marcel Worring. 2002. Document understanding for a broad class of documents. *IJDAR*, 5:1–16.

Antoine Bordes, Nicolas Usunier, and Jason Weston. 2010. Label ranking under ambiguous supervision for learning semantic correspondences. In *27th International Conference on Machine Learning (ICML 2010)*, pages 103–110.

Charles Elkan and Keith Noto. 2008. Learning classifiers from only positive and unlabeled data. In *Proceedings of the 14th ACM SIGKDD International Conference on Knowledge Discovery and Data Mining*, KDD '08, pages 213–220, New York, NY, USA. ACM.

Jing Fang, Zhi Tang, and Liangcai Gao. 2011. Reflowing-driven paragraph recognition for electronic books in PDF. In Gady Agam and Christian Viard-Gaudin, editors, *DRR*, volume 7874 of *SPIE Proceedings*, pages 1–10. SPIE.

Katja Filippova and Michael Strube. 2006. Using linguistically motivated features for paragraph boundary identification. In *Proceedings of the 2006 Conference on Empirical Methods in Natural Language Processing*, EMNLP '06, pages 267–274, Stroudsburg, PA, USA. Association for Computational Linguistics.

Christopher D. Manning, Prabhakar Raghavan, and Hinrich Schütze. 2008. *Introduction to Information Retrieval*. Cambridge University Press.

Timothy A Miller, Sean Finan, Dmitriy Dligach, and Guergana Savova. 2015. Robust sentence segmentation for clinical text. In *Proc AMIA Symp*, pages 112–113, San Francisco, Ca. AMIA.

Kamal Nigam, Andrew Kachites McCallum, Sebastian Thrun, and Tom Mitchell. 2000. Text classification from labeled and unlabeled documents using EM. *Machine Learning*, 39(2-3):103–134, mai.

Bogdan Radakovic, Sasa Galic, and Aleksandar Uzelac. 2013. Paragraph recognition in an optical character recognition (OCR) process. United States Patent 8,565,474 B2, US Patent Office, octobre.

Guergana K. Savova, James J. Masanz, and Philip V. Ogren. 2010. Mayo clinical Text Analysis and Knowledge Extraction System (cTAKES): architecture, component evaluation and applications. *Journal of the American Medical Informatics Association*, 17:507–513.

Noah A. Smith. 2011. *Linguistic Structure Prediction*. Synthesis Lectures on Human Language Technologies. Morgan and Claypool.

Caroline Sporleder and Mirella Lapata. 2006. Broad coverage paragraph segmentation across languages and domains. *ACM Trans. Speech Lang. Process.*, 3(2):1–35, juillet.

Amber Stubbs, Christopher Kotfila, Hua Xu, and Özlem Uzuner. 2015. Identifying risk factors for heart disease over time: Overview of 2014 i2b2/UTHealth shared task Track 2. *J Biomed Inform*.

M. Tepper, D. Capurro, F. Xia, L. Vanderwende, and M. Yetisgen-Yildiz. 2012. Statistical section segmentation in free-text clinical records. In *LREC 2012, Eigth International Conference on Language Resources and Evaluation*, Istanbul, Turkey. ELRA.

Guillaume Wisniewski, Nicolas Pécheux, Souhir Gahbiche-Braham, and François Yvon. 2014. Cross-lingual part-of-speech tagging through ambiguous learning. In *Conference on Empirical Methods in Natural Language Processing (EMNLP)*, pages 1779–1785, Doha, Qatar. Association for Computational Linguistics.

Pierre Zweigenbaum and Cyril Grouin. 2014. Reformatting clinical records based on global layout statistics. In Olivier Bodenreider, José Luis Oliveira, and Fabio Rinaldi, editors, *Proceedings 6th International Symposium for Semantic Mining in Biomedicine (SMBM 2014)*, pages 53–60, Aveiro. University of Aveiro.

# BioDCA Identifier: A System for Automatic Identification of Discourse Connective and Arguments from Biomedical Text


**Sindhuja Gopalan**
AU-KBC Research Centre
MIT Campus of Anna University
Chromepet, Chennai, India
sindhujagopalan@au-kbc.org

**Sobha Lalitha Devi**
AU-KBC Research Centre
MIT Campus of Anna University
Chromepet, Chennai, India
sobha@au-kbc.org


## Abstract


This paper describes a Natural language processing system developed for automatic identification of explicit connectives, its sense and arguments. Prior work has shown that the difference in usage of connectives across corpora affects the cross domain connective identification task negatively. Hence the development of domain specific discourse parser has become indispensable. Here, we present a corpus annotated with discourse relations on Medline abstracts. Kappa score is calculated to check the annotation quality of our corpus. The previous works on discourse analysis in bio-medical data have concentrated only on the identification of connectives and hence we have developed an end-end parser for connective and argument identification using Conditional Random Fields algorithm. The type and sub-type of the connective sense is also identified. The results obtained are encouraging.


## 1 Introduction

Due to advancements in bio-medical field, a large number of bio-medical literatures are available. It is crucial to extract knowledge from these literatures, to prevent the loss of important findings required for progression in bio-medical field. For extracting the information from the text, the text needs to be analyzed linguistically and it is absolutely essential to add such linguistic information to the text for future research. Natural language processing (NLP) methods are being widely used to analyse bio-medical text by performing tasks like automatic summarization, translation, named entity recognition, discourse analysis, speech recognition, etc.,. Discourse analysis is one such fundamental topic in NLP domain that makes a text linguistically rich. Discourse analysis is the study of the relation between phrase, clauses or sentences in a text. The basic units of discourse relations are discourse markers and their arguments. Discourse markers are words or phrases that establish a relation between two discourse units thereby connecting two events.

Example [1]

a) *Embryonic Stem cells have a high mitotic index and form colonies.* **So**, *experiments can be completed rapidly and easily.*

b) Clinical lung cancers containing a higher abundance of *ALDH* **and** *CD44* co expressing cells were associated with lower recurrence free survival.

In the above Example 1 (a) "so" is the discourse connective that indicates a relation between two discourse units. These discourse units are labeled as arguments. It is not easy to identify the discourse connectives, as all connectives are not discourse markers. In some cases it acts as conjunction simply unifying two words. Consider the Example 1 (b), where "and" acts as a conjunction connecting two bio-medical named entities "ALDH" and "CD44 co expressing cells".

The intended goal of our present work is to study the "discourse relation" in Medline abstracts and develop a system to automatically extract these relations. Cohen et al., (2010) has examined the structural and linguistic aspects of abstracts and bodies of full text articles. Their work shows that full parsing






*Proceedings of the Fifth Workshop on Building and Evaluating Resources for Biomedical Text Mining (BioTxtM 2016),*
pages 89–98, Osaka, Japan, December 12th 2016.

of article bodies is more difficult than abstracts as article bodies has longer sentences. They also assessed the incidence of conjunctions in abstracts and article bodies. The difference between abstracts and bodies was not statistically significant with only slightly more conjunction in article bodies than in abstracts. Hence, we used abstracts to develop the system instead of whole articles. Since abstracts give an overview of a work, the sentences in the abstract need to be coherent. Hence it is well connected by connectives. The occurrence and type of discourse connectives in the bio-medical domain vary from other domain.

Example [2]

*Furthermore, juglone blocked the adipogenic medium-induced activation of PPAR, C/EBP, C/EBP, and ERK pathways, which was rescued by Ad-PIN1 infection.* **In summary**, *the present study shows for the first time that PIN1 acts as a significant modulator of odontogenic and adipogenic differentiation of HDPSCs, and may have clinical implications for regenerative dentistry.*

In the above Example 2, the discourse marker "In summary" occurs commonly in bio-medical abstracts, whereas in general domain it is used to a lesser extent. There are many such connectives that occur in bio-medical domain and may or may not occur in other domains. Also, work done by Ramesh et al., (2012) show that the classifier trained on open domain performs poorly on bio-medical domain. Hence there is a demand to spring up a discourse parser for bio-medical domain.

Our work makes a substantial contribution to the development of the discourse annotated corpus for bio-medical domain. At this period, there is no end-end discourse parser available for automatically identifying the connectives, sense and its arguments in bio-medical domain; hence we have developed a discourse parser for explicit relation identification from bio-medical corpus. This system can further be used in NLP tasks like machine comprehension, extraction of semantic relations, co-reference resolution, etc. In the next section, works related to discourse analysis are detailed. Section 3 describes the annotation task. Section 4 explains the method used. Results are demonstrated and discussed in section 5. We conclude and outline our future work in section 6.

## 2   Related Works

In this section, we describe the annotation works and various systems developed for automatic identification of the discourse relations.

### 2.1   Annotation

Large scale annotated corpora for discourse analysis is developed in recent years that play a major role in natural language research. Lopatkova et al., (2009) in their project has assigned structure of clauses to Czech sentences from the Prague dependency tree, but as a new stratum of syntactic annotation. The PDTB adds low level discourse structure and semantics using connective specific semantic role labels (Prasad et al., 2008a). Dinesh et al., (2005) in their paper has disclosed substantial divergences among the syntactic structure and discourse structure in terms of the arguments of connectives. Miltsakaki et al., (2005) has presented a set of manual sense annotation studies for three connectives since, while and when. The work on discourse connectives also extends to other languages. The discourse relation has been studied in various languages like Turkish (Zeyrek and Webber , 2008), French (Roze et al., 2010), Arabic (Al-Saif et al., 2010), Hindi (Prasad et al., 2008b), Tamil (Rachakonda and Sharma, 2011; Menaka et al., 2011) and cross lingual variation is analyzed in Indian languages like Malayalam, Hindi and Tamil (Devi et al., 2014). Discourse annotation work has also been widened to bio-medical domain by Mihaila et al., (2013) and Tateisi et al., (2000)

The first work on the annotation of bio-medical discourse connectives and its arguments was done by Prasad et al., (2011). They have developed a Biomedical Discourse Relation Bank (BioDRB) in which they have annotated explicit and implicit discourse relation in 24 open access full text bio-medical articles from GENIA corpus containing 4911 sentences. They have adapted the annotation guidelines from PDTB, but have introduced new conventions and modification for sense classification. The related works show that BioDRB is the only discourse relation tagged corpus available in the bio-medical field. Motivated by the need to construct a corpus to be used for the text mining in the bio-medical field, we developed a discourse annotated corpus using Medline abstracts.

## 2.2 Systems Developed

Existing systems for connective and argument identification are developed based on the PDTB. Elwell and Baldridge (2008) shows that using models for specific connectives and the types of connectives and interpolating them with a general model improves performance. Lin et al., (2012) has developed a full discourse parser using the parsing algorithm in PDTB style. This parser first identifies all discourse and non-discourse relations, locates and labels their arguments, and then classifies their relation types. They reported overall system F-scores for partial matching of 46.80% with gold standard parser and 38.18% with full automation. Ghosh et al., (2011) in their work has taken a data driven approach to identify arguments of explicit discourse connectives using Conditional Random Fields (CRFs) technique and obtained an F-score of 57% for arg1 and 79% for arg2. Wellner et al., (2007) has worked on discovering the arguments of discourse connectives in the PDTB. Rather than identifying the full-extent of the arguments as annotated in the PDTB, they identified the argument heads. Using log-linear re-ranking model, they identified both the arguments correctly for over 74% of the connectives on held-out test data using gold standard parsers.

Stepanov and Riccardi (2014) in their paper has presented cross-domain evaluation of PDTB trained discourse relation parser and evaluated feature-level domain adaptation proficiencies on the argument span extraction sub task. They summed up that the corpora differences with respect to discourse connective usage affect the cross domain generalization of connective detection tasks negatively. Hence, it is necessary to develop a domain specific system for identification of discourse connectives. Ramesh et al., (2012) has developed a system for identification of discourse connectives in bio-medical domain and has obtained an F-score of 69%. But, they did not focus on identification of arguments. The main goal our work is to develop a system for automatic identification of explicit connectives, its sense and arguments using machine learning approach. The experiment outcomes manifested the efficiency of our system in discourse relation identification task.

## 3 Annotation Task

Our annotation task includes the annotation of discourse connectives, its sense and their arguments by following the guidelines of PDTB, a large-scale resource of annotated discourse relations and their arguments (Prasad et al., 2008a). We formulated our corpus by collecting abstracts from PubMed Central (PMC). PMC is a free full-text archive of bio-medical and life sciences journal literature developed by the U.S. National Institutes of Health's National Library of Medicine (NIH/NLM). To analyze the distribution of discourse connectives and its arguments, a corpus containing 7670 sentences is used. The bio-medical corpus is annotated with Explicit, Implicit, Altlex, NoRel and Entrel relations. The syntactic classification of connectives includes Subordinators, Coordinators, Conjunct adverbs, and Correlative conjunction.

Example [3]

*The practical application of ESCs is throttled*$_{\text{arg1}}$ **because** *it is unmanageable to derive and culture ESCs*$_{\text{arg2}}$.

In Example 3, the subordinator "because" occurs in the middle of the sentence connecting main clause with the subordinate clause. The sense of the discourse connectives provide a semantic description of the relation between the arguments of connectives. Based on the sense, these connectives are broadly classified into four top class levels viz Expansion, Temporal, Comparison and Contingency. These classes are further classified into various types and sub-types. For instance in Example 3, the connective "because" comes under sense class "Contingency" and belongs to the type "Cause" and sub-type "reason".

The arguments are labeled as arg1 and arg2. The arguments include single clauses or multiple clauses and in some cases it may include whole sentences and even multiple sentences. The clause or sentence that is syntactically attached to connective is labeled as arg2 and the other clause or sentence is marked as arg1. Furthermore the arguments may be adjacent or non adjacent to the connectives. According to the concept of minimality, the minimum required argument for a relation is annotated. There are no constraints for the linearity of the connectives and arguments. The linear order of the connectives and arguments need not be of the basic order arg1-con-arg2. It varies depending on the location of the connective. The connective can occur at sentence initial, medial or final position. In cases, where connective occurs in the initial position, the linear order is of the form con-arg2-arg1.

In case of implicit connectives, we have tagged the relation with a label "Implicit". Implicit relation can be inferred, where a relation exists between two discourse units but not explicitly marked.

Example [4]

*In adipogenic cultures, both cell populations showed positive Oil Red O staining by day 21*$_{arg1}$. **Implicit (Likewise, Similarly)** *in chondrogenic cultures, both stem cells expressed the formation of proteogly-can*$_{arg2}$.

The above Example 4 shows that there exists an implicit relation between two discourse units. The relation can be established by connectives like "likewise", "Similarly" etc.

AltLex relation is realized between adjacent sentences, where inserting an implicit connective may lead to redundancy in the expression of the relation.

Example [5]

*Dichaete potentially regulates many more genes in the Drosophila genome and was found to be associated with over 2000 mapped regulatory elements.* **AltLex [Our analysis suggests]** *that Dichaete acts as a transcriptional hub, controlling multiple regulatory pathways during CNS development.*

In the above Example 5 the AltLex relation is shown. Here, inserting a connective may cause redundancy. Hence AltLex relation is marked in such cases.

EntRel relation exists where the implicit relation between adjacent sentences is not between their abstract object interpretations, but form an entity based coherence. The entity is realized in both the sentences directly or indirectly.

Example [6]

*Brg1 is required for stem cell maintenance in the murine intestinal epithelium in a tissue-specific manner.* **EntRel** *Brg1 is a chromatin remodeling factor involved in mediation of a plethora of signaling pathways leading to its participation in various physiological processes both during development and in adult tissues.*

In Example 6, the entity "Brg1" in first discourse is directly realized in the second unit.

Norel relation exists where there is no discourse relation or entity based coherence relation between adjacent sentences. Since our corpus contains abstracts, all sentences within abstracts are related explicitly or implicitly or entity based coherence are found. However, in our corpus NOREL relation can be tagged between abstracts, as shown in Example 7.

Example [7]

*TGF-1 alone and in combination with PDGF also amplified surface integrin expression and adhesivity of MSCs with extracellular matrix proteins. These findings will provide a more mechanistic insight for modeling tissue-level rigidity in fibrotic tissues and tumors.* **NoRel** *Lung cancer tumorigenicity and drug resistance are maintained through ALDH(hi)CD44(hi) tumor initiating cells.*

The data statistics are shown in the Table 1.

| S.No | Connective Types | No of Connectives |
|------|------------------|-------------------|
| 1 | Explicit | 2957 |
| 1.a | Intra-sentential | 1742 |
| 1.b | Inter-sentential | 1215 |
| 2 | Implicit | 1610 |
| 3 | Altlex | 616 |
| 4 | EntRel | 802 |
| 5 | NoRel | 585 |

Table 1: Data statistics.

We annotated our corpus with above relations and cross checked the annotation quality by tagging

3000 sentences from the data using second annotator. We calculated the inter-annotator agreement using Cohens Kappa coefficient (Viera and Garrett, 2005), which is a statistical measure.

$$K = (p_o - p_c)/ (1 - p_c)$$

where $p_o$ is the agreement rate between two human annotators and $p_c$ is chance agreement between two annotators. The agreement between the annotators is almost perfect for connectives. We obtained Cohens kappa score of .94 for explicit connectives. In the case of annotation of arguments there is a substantial agreement between the annotators for all the argument boundaries. The overall agreement in identifying both the arguments of explicit connectives is 0.86. The variation in agreement rate is due to various structural inter-dependencies that occur between discourse relations.

## 4  Method

This section describes the experiments performed for extraction of discourse relations using machine learning (ML) approach. In this work we have concentrated on the automatic identification of explicit relations. We performed two sets of tasks. In the first task the connectives and its sense were identified and in the second task the argument boundaries were identified. In further sections experiments are explained in detail.

### 4.1  Features Used

For our work we have used simple and minimal number of features given in Table 2 and Table 3. The connectives are mostly conjunction and hence the PoS features contribute most to the identification of connectives. Chunk features help to identify the boundary of the connectives and arguments. Since a discourse connective connects two clauses, clause start and end can be used as feature for connective identification. For sense identification, we have used syntactic features. In addition to features used for connective identification, connective itself is used as a feature for identifying the sense of the connective.

| S.No | Features for Connectives | Examples |
|---|---|---|
| 1 | Word | previous word, current word, next word |
| 2 | PoS (P) | PoS of previous word, PoS of current word, PoS of next word |
| 3 | Combination of PoS and Chunk | PoS and Chunk of current word |
| 4 | Combination of 1, 2, 3 | Current word, PoS of current word, Chunk of current word |
| 5 | Connective | Connective is an exceptional feature for sense identification |

Table 2: Features used for connective identification (syntactic and sense).

| S.No | Features for Arguments | Examples |
|---|---|---|
| 1 | Sentence position with respect to connective | arg1 end mostly will be before connective and arg2 start after connective |
| 2 | Sentence boundary | arg1 start  mostly start of the sentence, arg2 end- mostly end of the sentence |
| 3 | Clause | Arguments may be clause and hence clause boundary is used as feature |
| 4 | Previously identified Argument boundaries | Previously identified arguments arg1 end, arg2 start, arg 1 start |

Table 3: Features used for argument identification.

## 4.2 Experiments Performed

The annotated corpus was preprocessed, before training it using ML algorithms. The sentences were tokenized and PoS tags and chunks were added using the GENIA tagger (Tsuruoka et al., 2005). After analyzing the corpus, features were extracted. Based on the extracted features, language models were built using CRFs algorithm. We used CRF++ tool (Kudo, 2005), an open source implementation of CRFs algorithm. Using the language model the explicit connectives, its sense and argument boundaries were automatically identified from the test set. The experiments were performed as two tasks.

**Connective classification and sense identification**: In the first task, the system was trained for classification of connectives. Using the features described in section 4.1, the model was built for connective identification. The built model was used for identifying the connectives from the test data. Further, post-processing rules were applied to improve the system's performance. After classifying the tokens as connectives and non-connectives, sense of the connectives were identified. In our work we have also identified the type and sub-type of the connective sense. We developed one-stage model and multi-stage model to identify the sense. For one stage model, a single model is developed for all types of sense. While for multi-stage model, four separate models were developed for each type of sense based on its upper level class. After identifying the senses separately, the output was combined from each model. The connectives having overlapping senses were merged based on confidence scores (i.e. sense having large probability).

**Argument identification**: After identifying the connectives, the second task was performed, where the arguments were identified. To overcome the problem of overlapping sequence, we processed each connective separately for argument identification. We developed two types of models, first by partitioning these sentences into inter and intra-sentential relations and second as a one-stage model without partitioning the sentences. Intra-sentential connectives are those that occur within a sentence, while inter-sentential connectives are those that occur outside the sentence. For inter-sentential connectives mostly previous sentence acts as argument. In few cases the arguments span across sentences. For the one-stage model the output from connective identification was given as such after extracting the features without dividing them as inter-sentential or intra-sentential connectives. We developed gold standard parser and automatic parser for argument identification using features mentioned in Table 3 .

For gold standard parser the gold standard connectives were used to train the system. For automatic parser the output from the connective identification task was fed as input to the argument identification task. For identifying the arguments we followed the method used by Menaka et al., (2011). They presented their work on automatic identification of the cause-effect relation from Tamil text. In their work they developed separate models for each boundary. Similarly, we built 4 models for each boundary of the arguments, i.e. identification of arg1 start and end and arg2 start and end. The argument boundaries were identified in the following series, arg2 start, arg1 end, arg1 start and arg2 end. The output from one model is fed as input to the next model. The choice of order of identification of bounds was made with the idea that it is easier to identify the boundaries that are close to the connective. After identifying the boundaries, the outputs were merged. Thus connective, sense and arguments were identified. The results are detailed in the next section.

## 5 Results and Discussion

We evaluated the performance of our system using precision, recall and F-score measure. Precision is the number of labels correctly perceived by the system from the total number of labels identified, Recall is the number of labels correctly detected by the system by the total number of labels contained in the stimulus text and F-score is merely the mean of precision and recall. We performed 10 fold cross-validations for connective identification using CRFs. The partition of the corpus was done randomly to test all the relations at least once. The average 10 fold cross-validation F-score obtained for CRFs is 86.42%. The result for connective classification obtained from best model is presented in Table 4. For sense identification the results for one-stage model and multi-stage model are presented in Table 5. For sense identification one-stage model gives better results than multi-stage model for both gold and automatic parser.

| Method | Precision (%) | Recall (%) | F-score (%) |
|---|---|---|---|
| CRFs | 92.49 | 83.83 | 88.16 |
| CRFs + Post-processing | 93.01 | 88.05 | 90.53 |

Table 4: Results for connective classification.

| Model | Gold Parser | | | Automatic Parser | | |
|---|---|---|---|---|---|---|
| | Precision | Recall | F-score | Precision | Recall | F-score |
| One-stage model | 98.03 | 93.26 | 95.65 | 90.79 | 79.25 | 85.02 |
| Multi-stage model | 97.79 | 91.22 | 94.51 | 90.87 | 78.13 | 84.5 |

Table 5: Results for sense identification in %.

Then the arguments of the connectives were identified using intra-sentential and inter-sentential model and also using one-stage model. It is observed that the F-score for identification of intra-sentential and inter-sentential argument boundaries arg1 end and arg2 start is better than arg1 start and arg2 end. This is because the argument boundaries arg1 end and arg2 start are nearer to discourse connective in intra-sentential model. While, in inter-sentential model the argument boundary arg1 end will be mostly the sentence or clause end and arg2 start will succeed the connective. The argument identification results for intra-sentential argument and inter-sentential arguments are shown in Table 6 and 7 respectively. The results for one-stage model is given in Table 8.

| Arguments | Gold Parser | | | Automatic Parser | | |
|---|---|---|---|---|---|---|
| | Precision | Recall | F-score | Precision | Recall | F-score |
| Arg1 start | 80.95 | 81.20 | 81.08 | 81.27 | 76.39 | 78.83 |
| Arg1 end | 94.11 | 89.24 | 91.68 | 84.31 | 81.2 | 82.76 |
| Arg2 start | 94.20 | 93.24 | 93.72 | 93.1 | 90.89 | 91.99 |
| Arg2 end | 85.31 | 80.54 | 82.93 | 83.3 | 81.56 | 81.93 |

Table 6: Results for intra-sentential argument identification in %.

The system achieved significant performance even with minimal features. The performance of existing systems described in Section 2.2, shows that the performance of our system is comparable to state-of-art systems. The errors in the identification of connectives and arguments were analysed. After identifying the errors we developed post-processing rules to improve the results. We analyzed the output obtained from the system and observed that the decrease in measures in automatic identification of connectives is due to data sparsity. The connective patterns that exist in test data may not exist in the training data. Also, the difficulty in identification of connectives arises due to propagation of errors from preprocessing modules. As we use PoS as a feature for connective identification the error introduced in PoS module decreases the measure. Importantly, conjunctions are not connectives.

Example [8]

*BMECs are important components of the hematopoietic microenvironment in the bone marrow*$_{arg1}$ **and** *they can secret several types of cytokines*$_{arg2}$.

In Example 1 (b) the conjunction "and" connect two noun "ALDH" and "CD44 co-expressing cells". Hence "and" does not act as a discourse connective. Whereas, in Example 8 "and" connects two discourse units as a whole and hence in this case it acts as a discourse connective. This ambiguity in the identification of connectives creates false positive results. To overcome this problem we have formulated some linguistic rules and have applied to the CRFs output. We explain an example for linguistic rule used in our work.

Rule1:

If the current token is "and", previous token is ",", PoS of previous to previous token is "VBN" and PoS of the next token is a "noun", then "and" is a discourse connective.

| Arguments | Gold Parser | | | Automatic Parser | | |
|---|---|---|---|---|---|---|
| | Precision | Recall | F-score | Precision | Recall | F-score |
| Arg1 start | 86 | 78 | 82 | 84 | 76.6 | 80.3 |
| Arg1 end | 86.5 | 83.4 | 84.9 | 85.6 | 82.4 | 84 |
| Arg2 start | 90.5 | 90.5 | 90.5 | 88 | 86.8 | 87.5 |
| Arg2 end | 82.1 | 81.5 | 81.8 | 80.5 | 76.3 | 78.3 |

Table 7: Results for inter-sentential argument identification in %.

| Arguments | Gold Parser | | | Automatic Parser | | |
|---|---|---|---|---|---|---|
| | Precision | Recall | F-score | Precision | Recall | F-score |
| Arg1 start | 82.9 | 80.1 | 81.5 | 83.4 | 76.8 | 80.1 |
| Arg1 end | 96.3 | 95.5 | 95.9 | 86.6 | 86.8 | 86.7 |
| Arg2 start | 99.4 | 98.2 | 98.8 | 92.1 | 85.4 | 88.8 |
| Arg2 end | 87.3 | 87.5 | 87.4 | 87.4 | 82.1 | 84.7 |

Table 8: Results for one-stage model argument identification in%.

Rule2:
If the current token is "also", PoS of previous token is "JJ" and PoS of the next token is "RB", then "also" is a disourse connective.

In the identification of arguments, the paired connectives generate errors. The paired connectives or co-relative conjunction includes two connectives that share same arguments. The error occurs in the argument boundary identification, when the argument includes multiple sentences. As we consider only the sentence with intra-sentential connective as input to intra-sentential model for argument identification the difficulty occurs in identification of argument with multiple sentences. Connectives can occur at the beginning or in the middle of the sentences or sometimes at the end. Consider below Example 9, where connective "also" occurs in the middle of the argument. This creates an error in identification of arg2 start.

Example [9]

*IL-17 activates several downstream signaling pathways including NF-B, MAPKs and C/EBPs to induce gene expression of antibacterial peptides, proinflammatory chemokines and cytokines and matrix metalloproteinases (MMPs)*$_{arg1}$. *IL-17 can **also** stabilize mRNAs of genes induced by TNF*$_{arg2}$.

The error analysis indicates the need for more sophisticated features to further improve the precision of the system.


## 6   Conclusion

We have developed a discourse relation bio-medical corpus, annotated with discourse connectives and its arguments. In this paper, we have explained the guidelines used for annotating our corpus. The inter-annotator agreement is calculated to check the annotation quality. We obtained almost perfect agreement between annotators for connectives and substantial agreement for argument boundaries. Also, a method for identification of discourse connectives and arguments from the annotated corpus using the ML approach is presented. We obtained encouraging results even with minimal features. In our future work, we will try to resolve the errors identified, thereby improving the overall results of the parser. Also, we will evaluate our system using BioDRB corpus, so that performance of our system on full articles can be verified. We will also extend our work in developing a discourse parser for identifying the implicit, EntRel and Altlex relations.

# References


Al-Saif, Amal, and Katja Markert. 2010. The Leeds Arabic Discourse Treebank: Annotating Discourse Connectives for Arabic. *Proceedings of the 6th International Conference on Language Resources and Evaluation.*

K Bretonnel Cohen, Helen L. Johnson, Karin Verspoor, Christophe Roeder and Lawrence .E Hunter 2010. The structural and content aspects of abstracts versus bodies of full text journal articles are different. *BMC Bioinformatics,* 11:492.

Anthony J. Viera and Joanne M. Garrett. 2005. Understanding Interobserver Agreement: The Kappa Statistic. *Research Series,* 37(5):360–363.

Balaji P. Ramesh, Rashmi Prasad, Tim Miller, Brian Harrington and Hong Yu. 2012. Automatic discourse connective detection in biomedical text. *J Am Med Inform Assoc,* 19(5):800–808.

Ben Wellner and James Pustejovsky. 2007. Automatically identifying the arguments of discourse connectives. *Proceedings of the Joint Conference on Empirical Methods in Natural Language Processing and Computational Natural Language Learning,* 92–101.

Charlotte Roze, Laurence Danlos and Philippe Muller. 2010. LEXCONN: a French lexicon of discourse connectives. *Proceedings of Multidisciplinary Approaches to Discourse,* Moissac, France, 114–125.

Claudiu Mihaila, Tomoko Ohta, Sampo Pyysalo and Sophia Ananiadou. 2013. BioCause: Annotating and analysing causality in the biomedical domain. *BMC Bioinformatics,* 14:2.

Deniz Zeyrek and Bonnie Webber. 2008. A Discourse Resource for Turkish: Annotating Discourse Connectives in the METU Corpus. *International Joint Conference on Natural Language Processing,* Hyderabad, India.

Eleni Miltsakaki, Nikhil Dinesh, Rashmi Prasad, Aravind Joshi and Bonnie Webber. 2005. Experiments on Sense Annotations and Sense Disambiguation of Discourse Connectives. *Proceedings of the 4th Workshop on Treebanks and Linguistic Theories,* Barcelona.

Evgeny A. Stepanov and Giuseppe Riccardi. 2014. Towards Cross-Domain PDTB-Style Discourse Parsing. *Proceedings of the 5th International Workshop on Health Text Mining and Information Analysis,* Gothenburg, Sweden, 30–37.

Marketa Lopatkova, Natalia Klyueva and Petr Homola. 2009. Capturing the Relationship among Clauses in Czech Sentences. *Proceedings of the Third Linguistic Annotation Workshop,* Suntec, Singapore, 74–81.

S Menaka, Pattabhi RK. Rao and Sobha L. Devi. 2011. Automatic identification of cause-effect relations in Tamil using CRFs. *Proceedings of Computational Linguistics and Intelligent Text Processing,* Springer, Berlin, Heidelberg, 316–327.

Nikhil Dinesh, Alan Lee, Eleni Miltsakaki, Rashmi Prasad, Aravind Joshi and Bonnie Webber. 2005. Attribution and the (Non-)Alignment of Syntactic and Discourse Arguments of Connectives. *Proceeding Corpus Anno '05 Proceedings of the Workshop on Frontiers in Corpus Annotations II: Pie in the Sky,* Stroudsburg, PA, USA, 29–36.

Ravi T. Rachakonda, and Dipti M. Sharma. 2011. Creating an Annotated Tamil Corpus as a Discourse Resource. *Proceedings of the Fifth Law Workshop (LAW V),* Portland, Oregon, 119-123.

Rashmi Prasad, Nikhil Dinesh, Alan Lee, Eleni Miltsakaki, Livio Robaldo, Aravind Joshi and Bonnie Webber. 2008a. The Penn discourse Treebank 2.0. *Proceedings of the 6th International Conference on Language Resources and Evaluation.*

Rashmi Prasad, Husain S, Dipti M. Sharma and Aravind Joshi. 2008b. Towards an Annotated Corpus of Discourse Relations in Hindi. *International Joint Conference on Natural Language Processing.* Hyderabad, India.

Rashmi Prasad, Susan McRoy, Nadya Frid, Aravind Joshi and Hong Yu. 2011. The biomedical discourse relation bank. *BMC Bioinformatics,* 12:188.

Robert Elwell and Jason Baldridge. 2008. Discourse connective argument identification with connective specific rankers. *Proceedings of the IEEE International Conference on Semantic Computing,* Washington, DC, USA, 198–205.

Sobha L. Devi, Lakshmi Sreedhar and Sindhuja Gopalan. 2014. Discourse Tagging for Indian Languages. *Lecture Notes in Computer Science,* 8404:470–481.

Sucheta Ghosh, Richard Johansson, Giuseppe Riccardi, and Sara Tonelli. 2011. Shallow Discourse Parsing with Conditional Random Fields. *Proceedings of the 5th International Joint Conference on Natural Language Processing*, Chiang Mai,Thailand, 1071–1079.

Taku Kudo. 2005. CRF++, an open source toolkit for CRF. *http://crfpp.sourceforge.net.*

Yoshimasa Tsuruoka, Yuka Tateishi, Jin-Dong Kim, Tomoko Ohta, John McNaught, Sophia Ananiadou, and Jun'ichi Tsujii. 2005. Developing a Robust Part-of-Speech Tagger for Biomedical Text. *Lecture Notes in Computer Science*, 3746:382–392.

Yuka Tateisi, Tomoko Ohta, Nigel Collier, Chikashi Nobata and Jun-ichi Tsujii. 2000. Building an Annotated Corpus in the Molecular-Biology Domain. *Proceedings of the COLING-2000 Workshop on Semantic Annotation and Intelligent Content*, 28–36.

Ziheng Lin, Hwee Tou Ng and Min-Yen Kan. 2012. A PDTB-Styled End-to-End Discourse Parser. *Natural Language Engineering*, 1(1):1–35.

# Data, tools and resources for mining social media drug chatter


**Abeed Sarker**
Division of Informatics
Department of Biostatistics and Epidemiology
The Perelman School of Medicine
University of Pennsylvania
abeed@upenn.edu

**Graciela Gonzalez**
Division of Informatics
Department of Biostatistics and Epidemiology
The Perelman School of Medicine
University of Pennsylvania
gragon@upenn.edu



## Abstract

Social media has emerged into a crucial resource for obtaining population-based signals for various public health monitoring and surveillance tasks, such as pharmacovigilance. There is an abundance of knowledge hidden within social media data, and the volume is growing. Drug-related chatter on social media can include user-generated information that can provide insights into public health problems such as abuse, adverse reactions, long-term effects, and multi-drug interactions. Our objective in this paper is to present to the biomedical natural language processing, data science, and public health communities data sets (annotated and unannotated), tools and resources that we have collected and created from social media. The data we present was collected from Twitter using the generic and brand names of drugs as keywords, along with their common misspellings. Following the collection of the data, annotation guidelines were created over several iterations, which detail important aspects of social media data annotation and can be used by future researchers for developing similar data sets. The annotation guidelines were followed to prepare data sets for text classification, information extraction and normalization. In this paper, we discuss the preparation of these guidelines, outline the data sets prepared, and present an overview of our state-of-the-art systems for data collection, supervised classification, and information extraction. In addition to the development of supervised systems for classification and extraction, we developed and released unlabeled data and language models. We discuss the potential uses of these language models in data mining and the large volumes of unlabeled data from which they were generated. We believe that the summaries and repositories we present here of our data, annotation guidelines, models, and tools will be beneficial to the research community as a single-point entry for all these resources, and will promote further research in this area.


**Keywords:**

Social media, data mining, public health, natural language processing, data science.

## 1   Introduction

In recent years, social media has become a crucial platform for communication, discovery of information, and the sharing of opinions and views [1]. Thus, social media has also emerged as a resource for collecting real-time data directly from public discussions. The social media sphere continues to grow [2], and websites like Twitter attract significant numbers of daily users. Twitter currently has 289,000,000 active users with the number of registered users rising by 135,000 every day [3]. With 58 million tweets per day (9,100 tweets per second), Twitter data is content-rich on everyday




*Proceedings of the Fifth Workshop on Building and Evaluating Resources for Biomedical Text Mining (BioTxtM 2016),*
pages 99–107, Osaka, Japan, December 12th 2016.

discussions. As a result, Twitter, in addition to other popular social networks, is being actively utilized for a range of tasks including advertising [4], opinion mining [5], political analytics [6], and public health monitoring [7].

From the perspective of public health, systems have been proposed for a variety of tasks including the tracking of the spread of infectious diseases [8, 9], monitoring of prescription and illicit drug abuse [10-12], pharmacovigilance [13], and the monitoring smoking patterns [14]. Despite the obvious use cases for utilizing social media data, national surveillance programs are yet to integrate proposed systems [2]. A prime reason for this are the numerous challenges associated with the use of social media data. While early, keyword-based systems were easily deployable [15], their shortcomings have also been identified [16]. Solving complex natural language processing problems with social media data introduce additional challenges—such as dealing with the use of colloquial language and misspellings [17]. Even data collection from social media faces challenges due to these factors. In addition, the notoriously noisy nature of social media data, and data imbalance hinder system performances [13]. As a result, despite the abundance of health-related knowledge that is encapsulated within the vast social media domain, it is still significantly under-utilized in practical systems.

## 1.1  Social media and data science

Over the last several years, a flurry of research tasks has successfully employed supervised learning systems that use manually annotated data to solve various natural language processing (NLP) problems. These include, for example, text classification tasks such as detecting mentions of adverse drug reactions [22], and extracting exact mentions using sequence labeling techniques [23]. While these approaches have shown good performance in noisy, social media text, their need for manual annotations make them expensive in nature. Manual annotations are time consuming, and the erratic properties of social media text make annotation tasks even harder. Consequently, even designing annotation tasks and guidelines require significant amounts of expert time, experience in annotations, and exposure to user posted texts. While research from the recent past [13] has elaborated the need for data annotation efforts, the importance of developing standardized annotation guidelines for social media based non-standard data sets have been somewhat overlooked. Therefore, in addition to the need for publicly available targeted data and models, there is also a need for the development of social media text annotation guidelines that to ensure consistency in annotation standards.

The majority of the data available from social media is unlabeled. Recent advances in NLP has seen the effective application of language models learned from large volumes of unlabeled data for various text mining tasks. While the ability to learn language models from large data sets presents new possibilities, social media oriented public health monitoring research has still not actively applied these techniques. One reason behind this is that targeted data from social media for specific public health monitoring tasks is still scarce. Thus, there is a strong motivation for the public release of such data and models. For example, recent approaches for generating distributed word representations [18-20] from large, unstructured data sets have seen growing popularity. However, availability of such language models learned from relevant social media data is limited.

## 1.2  Aims

We have several aims for this broad coverage paper. These aims are summarized as follows:

1. To outline our annotated data and the resources we have created over the last several years, as part of a National Institute of Health (NIH) research grant [21] on mining social media for discovering adverse drug reactions.
2. To make available our evolving social media text annotation guidelines for pharmacovigilance and toxicovigilance so that these annotation guidelines can be followed for future annotation tasks.
3. To provide insights about our annotated corpora, annotation tasks, unlabeled data and models.
4. To discuss some of the utilities of our data sets and their potential future uses.

The rest of the paper is organized as follows. In the next section we present (i) our data collection technique, which expands on keyword-based approaches to include common, phonetically similar misspellings of drug names, (ii) our preparation of various publicly available annotated data sets, (iii) our detailed annotation guideline preparation, and (iv) our language model generation techniques. In the *Discussion* section we present some statistics and utilities of our published resources and tools, including potential applications of our unlabeled data and language models.

## 2    Methods

### 2.1    Data collection

Prior to collecting data, we selected a set of drugs of interest, which were likely to have a large number of associated comments in social media. In particular, we selected drugs that were prescribed for chronic diseases and syndromes for which large numbers of comments were expected and drugs with high prevalence of use (as per the IMS Health's Top 100 drugs by volume for the year 2013 [22]). Starting with this initial list of drugs, we added various drug names based on interest since 2014, such as drugs that may be prone to abuse. The final drug list is monitored by our in-house pharmacology expert, and further details about the drugs can be found in our past publications [22,23].

We collected data from Twitter using the drug names (trade and generic) as keywords. To address the issue of misspelled drug names, which affects recall during data collection, we developed a spelling variant generator [24]. The generator first identifies lexically close misspellings, specifically those that are 1-edit distance away in terms of Levenshtein distance. Phonetically similar misspellings are then identified, and finally, the Google custom search API is used to identify a smaller set of misspellings that are commonly used by users. We have made a downloadable version of our generator publicly available.[1] The generator is semi-automatic. Figure 1 presents a random sample of tweets associated with a number of drugs that were collected using our technique. The tweets appear to present a number of types of information, such as symptoms/indications, perceived adverse drug reactions, medication abuse information, user sentiments towards drugs and/or prices, and potential drug abuse, to name a few. The figure also illustrates how some drug names are often misspelled. Depending on the intent, distinct types of drug-related information can be mined from this data source.

### 2.2    Data annotation, guidelines and resources

Following the collection of large amounts of drug-related chatter from Twitter, we allocated significant resources to perform annotation of the data and for the preparation of standardized annotation guidelines. The annotation guidelines were prepared in consultation between experienced language annotators, NLP experts, public health professionals, and a pharmacology expert. The guidelines were finalized by the pharmacology expert after multiple iterations. The annotation guidelines also evolved over time, which is a necessity for social media data, as new characteristics of the data were discovered during the early iterations of annotation. Using the annotation guidelines, we were able to achieve high inter annotator agreements for our various annotation tasks. For adverse drug reaction detection from social media, we first performed binary annotations indicating if user posts mentioning at least one drug mentioned an adverse reaction or not (inter annotator agreement $\kappa = 0.74$). Following that, we performed annotations to tag specific mentions of adverse reactions and indications ($\kappa = 0.81$), including mapping the mentions to standardized IDs in the Unified Medical Language System (UMLS) vocabulary. We have made these detailed annotation guidelines publicly available to support future annotation tasks.[2] In addition to the guidelines, we have made resources associated with our classification and extraction tasks publicly available [22,23]. These include source codes, executable applications, lexicons, topics, cue words, word clusters, word embeddings, and annotated data, which we discuss later.

---

[1] Available at: http://diego.asu.edu/Publications/ADRSpell/ADRSpell.html. Last accessed: 2nd October, 2016.
[2] Available at: http://diego.asu.edu/guidelines/adr_guidelines.pdf. Last accessed: 2nd October, 2016.

can't sleep, **temazepam** myself into a coma, pass out for hours on end. finally wake up, feel like shite for days. Oh I love my life! :-/

my fibromyalga is killing me lately. has anything worked for u? **lyrica** and **neurontin** f'd up my life. **cymbalta** worse

just got retested for jcv. **tecfidera** did not work out well for me, so i'm onto **tysabri**. #ms #multiplesclerosis

**adderal** made me manic, **saphris** makes my skin crawl and gives me the dreaded twitches, **hydroxyzine** is more like a placebo than anything else

list of psychiatric medications i take for various psychiatric reasons. !. **saphris**. 2. **lamictal**. 3. **hydroxygine**. 4. **trazodone**. 5. **zoloft**.

the only kind i have is sleeping **siroquil** and it knocks me out for too long to make it to class

the sun is up & i haven't slept yet! the **quetiapine** is not knocking me out like it used to. been up for 24 hours & i aint sleepy :-( #bipol

snorted 2 15mg **oxycodone** ($24)

also **adderall** prevents me from having any feelings other than tired rage

i hate how this firbo and **gabapentin** robs me if my life ... i just hate feeling so useless and worthless feeling tired

i am taking a cocktail of **tramadol, acoxia, myonol** & **pregabalin** twice a day and i still cannot control this pain. huhuhuhuhuh

do not take **victoza** if you are allergic to **victoza** ... i an now worried about people who actually need this warning

i'm trying to go off it. i'm on **lamictal** now and it works but i'm still addicted to **Geodon**

my memory is still so awful, hate the side effects of **pregabalin** -.-

**Figure 1.** Sample tweets containing drug names including some that are misspelled, but were caught our common misspelling generator. The tweets present a variety of different types of information including the symptoms effectiveness of drugs, adverse reactions, user sentiments, and potential abuse of prescription medications.

In addition to our work on pharmacovigilance, our experts have collaborated to create guidelines and resources for additional tasks such as prescription medication abuse monitoring from social media. Similar to our other tasks, the annotations were carried out in several iterations and the guidelines prepared have been made publicly available.[3] We have also made some of our annotated research data on peripheral topics, such as prescription medication abuse, publicly available.[4]

As discussed in the abovementioned guidelines, annotation of social media data presents a variety of challenges, which must be addressed in a consistent manner. For Twitter, the first challenge faced when performing binary annotations was the lack of context. Due to the character limit of 140 per post, even for human annotators, it is often difficult to determine the context in which a potential adverse reaction is mentioned or if a mentioned adverse reaction represents a personal experience or just a general statement. Other factors, such as posts that are spread over multiple tweets also add to this problem. To address these and other annotation difficulties, regular meetings were held between the annotators and the pharmacology expert, during which common difficult annotation issues were identified, discussed, and resolved. We provide further details of common social media text annotation problems that we faced in the *Discussion* section.

### 2.3    Unlabeled data and language models

Besides preparing and releasing the largest annotated social media data sets for pharmacovigilance and other tasks, we also released unlabeled data and language models derived from the data. Language models generated from unstructured data sets, such as those via deep learning techniques, have recently received significant research attention because of their ability to capture semantic information [18]. We released two sets of language models for the research community, along with the data (approximately a quarter million tweets) used to create the models.[5] The following is a brief overview of each set.

The first set of models were prepared using the word2vec tool,[6] and they capture distributional and semantic information. Phrases/terms are represented using vectors using these models, with the vector sizes largely determining where each phrase appears in semantic space. We generated models with vector sizes between the sizes 200 and 400. For the different vector sizes, we generated models using context windows within the range [2,9]. Such distributed word representation models are already being applied for research utilizing other sources of noisy health-related data, such as clinical reports [25]. Our second set of models are sequential, and these language models capture the probabilities of n-gram sequences. These models have been applied for a variety of tasks in the past, such as lexical normalization [26]. In a sequential language model, the conditional probability of a term given all the previous terms is given as $P(t_1^M) = \prod_{k=1}^{M} P(t_k|t_1^{k-1})$, where $t_k$ is the $k^{th}$ term. To generate the n-gram language models, we used the KenLM n-gram, language modeling tool [27]. We have also made available a set of n-gram language models (n= 2—4) from the same unlabeled data set.

## 3    Discussions

In this section, we briefly discuss some of the uses of the various resources that we have published. The value of most of our various annotated data sets has already been established, and there has been a sizable amount of recent research that have utilized these data sets for tasks such as classification and extraction. The resources associated with our annotated data sets, such as the lexicons, word clusters, and so on, have been used for research outside the domain of pharmacovigilance. For binary classification of adverse drug reaction classification, we currently have a total of 25,678 annotated posts, which were prepared in 3 batches. 10,822 posts were made publicly available with our system/source code for social media text classification for pharmacovigilance [22].[7] Additional data sets

---


[3] Available at: http://diego.asu.edu/guidelines/DrugAbuseAnnotationGuideline1.1.pdf. Accessed 2nd October, 2016.

[4] Available at: http://diego.asu.edu/Publications/DrugAbuse_DrugSafety.html. Accessed 2nd October, 2016.

[5] Available at: http://diego.asu.edu/Publications/Drugchatter.html. Accessed 2nd October, 2016.

[6] https://www.tensorflow.org/versions/r0.11/tutorials/word2vec/index.html. Accessed: 20th October, 2016.

[7] Resources, tools and data are available at: http://diego.asu.edu/Publications/ADRClassify.html. Accessed: 26th October, 2016.

were made available to the participants of a shared task that we organized [33], and these data sets will also be made available via the link mentioned above. For adverse drug reaction mention extraction, we have made available 1784 annotated posts publicly available along with our state-of-the-art extraction system [23].[8] In total, we have 2607 annotations for this task, with the rest of the data only available to our shared task participants and will be made publicly available in the near future. We have also made available a collection of resources for social media mining for pharmacovigilance along with our review of the domain [13].[9]

Annotating biomedical data or social media data are challenging tasks and require expertise with the domains. The challenges are exacerbated when it comes to biomedical data from social media. As mentioned in the *Methods* section, we faced several frequently occurring annotation difficulties, which had to be resolved via multiple meetings and paired annotation sessions. The lack of context available with the short Twitter posts often made it difficult to determine if a post mentioned a personal experience of adverse reaction or just mentioned an adverse reaction for other reasons (*e.g.*, in many posts we found users simply repeating adverse reactions mentioned in television commercials). In many cases, our annotators found it difficult to determine if a mentioned condition was an adverse reaction or a symptom for which the drug in question was taken. Annotating the spans of concept mentions is even more challenging. Non-standard expressions (*e.g.*, '*head feels like a zombie*') and disjoint mentions of adverse reactions (*e.g.*, '*gives me pain in my freakin stomach*') are two of the leading causes of these difficulties. In addition to annotating the spans, our annotators were also required to map them to IDs in the UMLS. Non-standard adverse reaction mentions and context ambiguities led to numerous cases where more than one concept ID seemed valid. To resolve difficulties in selecting concept IDs, our annotators used paired annotation to identify IDs that were the most concrete fits, and developed specific, step-wise rules which are detailed in the previously mentioned annotation guideline.

Because of the costs and difficulties faced when annotating data within this complex domain, the preparation of comprehensive guidelines, such as ours, is of paramount importance. Detailed annotation guidelines with specific examples of problem cases can significantly reduce time required to plan for and design annotation tasks for social media based NLP studies. Even within the same annotation task there are inconsistencies in distinct research groups. We discovered such inconsistencies, for example, in the several data sets for binary classification of adverse drug reaction mentions. Therefore, we believe that our publicly available annotation guidelines will be helpful for the better understanding of potential issues associated with annotation of social media health data and to plan future annotation tasks.

We have discussed the recent release of a small batch of unlabeled data and sets of language models that were prepared using this data [32]. Analysis of that batch of unlabeled data revealed that discussions associated with drugs are generally skewed in Twitter, with some drugs discussed much more frequently than others. In the abovementioned sample of unlabeled data, while the distribution of tweets over the months were similar, we found some drugs to have a very large number of tweets associated with them. Figure 2 illustrates this information, showing that among the discussions regarding the top 10 most discussed drugs, 56% of the discussion was about Adderall® and 12% was about Xanax®. We suspect that the skewness in the distribution of drugs in social media chatter is because of the demographics among which social media is popular. Adderall®, for example, is a popular medication for abuse among young students, and, therefore, there is a large amount of chatter available for this drug, particularly during typical college examination times (*e.g.*, November/December) [10,28].

---

[8] Resources, tools and data are available at: http://diego.asu.edu/Publications/ADRMine.html. Accessed: 26th October, 2016.

[9] Available at: http://diego.asu.edu/Publications/ADRSMReview/ADRSMReview.html. Accessed: 26th October, 2016.

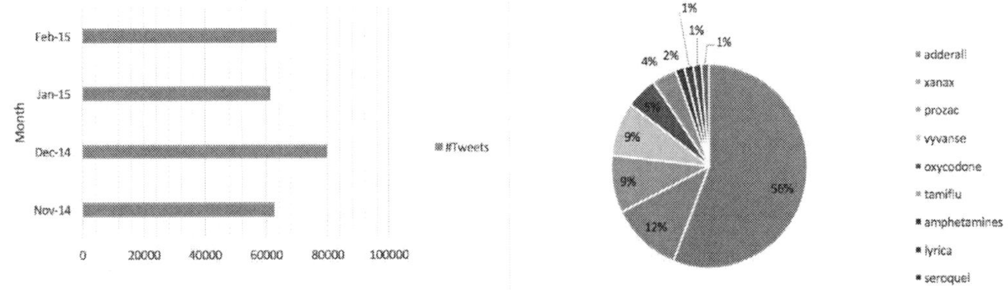

**Figure 2.** Distribution of drug related tweets over time and over different drug related keywords.

Past research has explored co-occurrence based techniques for identifying drug-adverse reaction associations [29]. One of the properties of the distributional semantics model is the ability to capture semantic associations between terms based on co-occurrence, and we have performed preliminary experimentation to assess the use of our models for drug-adverse reaction association identification. For the drug Trazodone, using one of our distributed representation models with a vector size of 400 and context window size 9, we compared the cosine similarity values between a drug keyword and a set of adverse reaction terms. Our similarity computations produced relatively high scores for known adverse reactions (the first four reactions from the left in Figure 3) and low scores for reactions for which no associations are known. While the threshold for this drug appears to be between 0.3 and 0.4, we could not establish specific values during our preliminary experimentation. Experimentation with other drugs (*e.g.*, such as those presented in [32]), also suggest that thresholds may vary between drugs or classes of drugs. Furthermore, there are unsolved NLP based problems, such as the vector representation of multi-word adverse reaction expressions. We also performed preliminary experiments with our sequential language models, such as assessing their usage in text classification. Because our data set essentially consists of health-related tweets, we used the sequential models to score a sample of posts from a separate data set containing annotations for health related tweets [31]. We observed that in general, health-related posts obtained higher scores compared to non-health related posts, as was expected. However, as with the distributional language models, we could not identify thresholds in the preliminary experimentation. We plan to address some of these limitations of our work in future research. We believe that incorporation of information from these models will improve the existing tasks of classification and extraction, and will be crucial for previously unexplored tasks such as concept normalization.

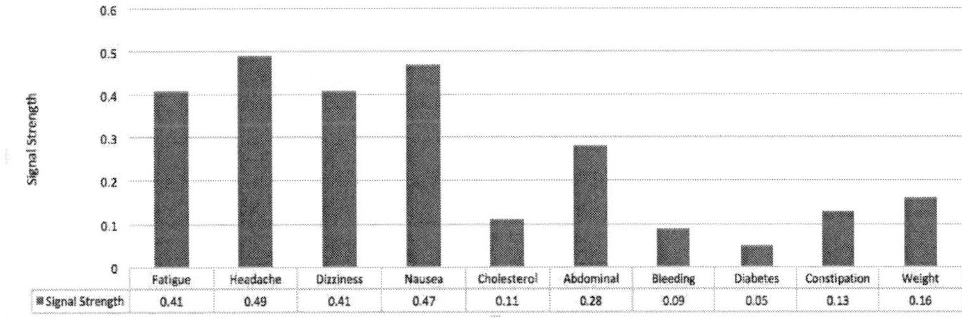

**Figure 3.** Association between trazodone and 10 adverse reactions computed using the distributed language models and cosine similarity.

The experimental results obtained from the use of our language models are very promising. With very simplistic settings, there appears to be a clear use case for these models for the tasks discussed. Our planned future work involves in-depth exploration of the various parameters of these models (*e.g.*, effect of context window sizes). We also encourage the research community to investigate the properties of the distinct models we are making available, and their applications. As discussed earlier, studies have already focused on extracting drug abuse information from social media, assessing the safety of drugs, exploring the prevalence of use of drugs, and discovering user sentiments towards specific drugs, to name a few. The linguistic regularities and the semantic knowledge captured by these models are likely to be useful for a number of important research tasks.

With the ever growing size of social media data, and the development of more efficient data processing techniques, the broader health domain will invariably benefit from utilizing social media data. However, it has also been realized that the *right* data is more important than *big* data, and the development of effective systems benefit from access to the former. Therefore, we believe that our released data, tools and resources, which have been summarized in this paper, will be very useful to the research community.

## References


1. Horvitz E, Mulligan D. Data, privacy, and the greater good. Science 2015; 349 (6245):253—255. PMID: 26185242.

2. Velasco E, Agheneza T, Denecke G, et al. Social media and internet-based data in global systems for public health surveillance: a systematic review. Milbank Q 2014; 92 (1): 7—33. PMID: 24597553.

3. Web reference. WebCite archive: http://www.webcitation.org/6ceI98x1Z. Original URL: http://www.statisticbrain.com/twitter-statistics/.

4. Khang H, Ki E-J, Ye L. Social Media Research in Advertising, Communication, Marketing, and Public Relations, 1997—2010. Journal Mass Commun 2012; 89 (2): 279—298. DOI 10.1177/1077699012439853.

5. Hu M, Liu B. Mining and summarizing customer review. Proceedings of the 10th ACM SIGKDD international conference on Knowledge Discovery and Data Mining; 2004 Aug 22—25; Seattle, Washington. 168—177. ACM; 2004.

6. Eom Y-H, Puliga M, Smailović et al. Twitter-Based Analysis of the Dynamics of Collective Attention to Political Parties. PLoS One 2015; 10 (7). DOI 10.1371/journal.pone.0131184.

7. Denecke K, Krieck M, Otrusina L et al. How to Exploit Twitter for Public Health Monitoring. Methods Inf Med 2013; 52 (4): 326—339. PMID: 23877537.

8. Paul MJ, Dredze M. You Are What You Tweet: Analyzing Twitter for Public Health. Proceedings of the International AAAI Conference on Weblogs and Social Media: 2011 July 17—21; Barcelona, Spain. AAAI Press; 2011.

9. Broniatowski DA, Paul MJ, Dredze M. National and Local Influenza Surveillance through Twitter: An Analysis of the 2012-2013 Influenza Epidemic. PLoS One 2013; 8 (12): e84672. PMID: 24349542.

10. Hanson CL, Burton SH, Giraud-Carrier C et al. Tweaking and tweeting: Twitter for nonmedical use of psychostimulant drug Adderall among college students. J Med Internet Res 2013; 15 (4): e62. PMID: 23594933.

11. Hanson CL, Cannon B, Burton S et al. An Exploration of Social Circles and Prescription Drug Abuse through Twitter. J Med Internet Res 2013; 15 (9): e189. PMID: 24014109.

12. Cavazos-Rehg, Krauss M, Fisher SL et al. Twitter chatter about marijuana. J Adolesc Heal 2015; 56 (2): 139—145. PMID: 25620299.

13. Sarker A, Ginn R, Nikfarjam A et al. Utilizing social media data for pharmacovigilance: A review. J Biomed Inform 2015; 54: 202—212. PMID: 25720841.

14. Struik LL, Baskerville NB. The role of Facebook in Crush the Crave, a mobile- and social media-based smoking cessation intervention: qualitative framework analysis for posts. J Med Internet Res 2014; 16 (7): e170. PMID: 25016998.

15. Cook S, Conrad C, Fowlkes A et al. Assessing Google Flu Trends Performance in the United States during the 2009 Influenza Virus A (H1N1) Pandemic. PLoS One 2011; 6 (8): e23610. PMID: 21886802.

16. Lazer D, Kennedy R, King G et al. The Parable of Google Flu: Traps in Big Data Analysis. Science 2014; 343 (6176): 1203—1205. PMID: 24626916.

17. Leaman R, Wojtulewicz L, Sullivan R et al. Towards Internet-Age Pharmacovigilance: Extracting Adverse Drug Reaction from User Posts to Health-Related Social Networks. Proceedings of the 2010 Workshop on Biomedical Natural Language Processing: 2010 July 15; Uppsala, Sweden. 117—125. Association for Computational Linguistics; 2010.

18. Mikolov T, Chen K, Corrado G et al. Efficient Estimation of Word Representations in Vector Spaces. Proceedings of the Workshop at the International Conference on Learning Representations: 2014 May 2—4; Scottsdale, Arizona. Archived at: arXiv:1312.5650v3.

19. Mikolov T, Sutskever I, Chen G et al. Distributed Representations of Words and Phrases and their Compositionality. Proceedings of the Twenty-seventh Annual Conference on Neural Information Processing Systems: 2013 December 5—10; Lake Tahoe, Nevada. Curran Associates, Inc; 2013.

20. Mikolov T, Yih W, Zweig G. Linguistic Regularities in Continuous Space Word Representations. Proceedings of NAACL-HLT: 2013 Jun 9—14; Atlanta, Georgia. 746—751. Association for Computational Linguistics; 2013.

21. NIH Grant Number 5R01LM011176, Mining Social Network Postings for Mentions of Potential Adverse Drug Reactions. 2012. RePORT URL: https://projectreporter.nih.gov/project_info_description.cfm?projectnumber=5R01LM011176-02.

22. Sarker A, Gonzalez G. Portable automatic text classification for adverse drug reaction detection via multi-corpus training. J Biomed Inform 2015; 53: 196—207. PMID: 25451103.

23. Nikfarjam A, Sarker A, O'Connor K et al. Pharmacovigilance from social media: mining adverse drug reaction mentions using sequence labeling with word embedding cluster features. J Am Med Informatics Assoc 2015; 22 (3): 671—681. PMID: PMID: 25755127.

24. Pimpalkhute P, Patki A, Nikfarjam A. Phonetic Spelling Filter for Keyword Selection in Drug Mention Mining from Social Media. AMIA Jt Summits Transl Sci Proc 2014; 2014: 90—95. PMID: 25717407.

25. Henriksson A, Kvist M, Dalianis H et al. Identifying adverse drug event information in clinical notes with distributional semantic representations of context. J Biomed Inform 2015; 57: 333—349. PMID: 26291578.

26. Han B, Baldwin T. Lexical normalization of short text messages: makn sens a #twitter. Proceedings of ACL-HLT: 2011 Jun 19—24; Portland, Oregon. 368—378. Association for Computational Linguistics; 2011.

27. Heafield K, Pouzyrevsky I, Clark JH et al. Scalable Modified Kneser-Ney Language Model Estimation. Proceedings of ACL: 2013 Aug 4—9; Sofia, Bulgaria. 690—696. Association for Computational Linguistics; 2013.

28. Sarker A, O'Connor K, Ginn R, Scotch M, Smith K, Malone D, Gonzalez G. Social Media Mining for Toxicovigilance: Automatic Monitoring of Prescription Medication Abuse from Twitter. Drug Saf. 2016 Mar;39(3):231-40. doi: 10.1007/s40264-015-0379-4. PMID: 26748505.

29. Nikfarjam A, Gonzalez G. Pattern mining for extraction of mentions of Adverse Drug Reactions from user comments. AMIA Annu Symp Proc 2011; 2011: 1019—1026. PMID: 22195162.

30. Toutanova K, Moore RC. Pronunciation modeling for improved spelling correction. Proceedings of ACL: 2002 Jul 7—12; Philadelphia, Pennsylvania. 144—151. DOI: 10.3115/1073083.1073109. Association for Computational Linguistics; 2002.

31. Paul MJ, Dredze M. A Model for Mining Public Health Topics from Twitter. Technical Report. Johns Hopkins University 2011. Archived at: http://www.cs.jhu.edu/~mpaul/files/2011.tech.twitter_health.pdf.

32. Sarker A, Gonzalez G. A corpus for mining drug-related knowledge from Twitter chatter: Language models and their utilities. [To Appear]. Data Brief. 2016.

33. Sarker A, Nikfarjam A, Gonzalez G. Social Media Mining Shared Task Workshop. Pac Symp Biocomput. 2016;21:581—592. PMID: 26776221.

# Detection of Text Reuse in French Medical Corpora


**Eva D'hondt**
LIMSI, CNRS
Université Paris-Saclay
F-91405 Orsay
dhondt@limsi.fr

**Cyril Grouin**
LIMSI, CNRS
Université Paris-Saclay
F-91405 Orsay
grouin@limsi.fr

**Aurélie Névéol**
LIMSI, CNRS
Université Paris-Saclay
F-91405 Orsay
neveol@limsi.fr

**Efstathios Stamatatos**
Dept. of Information and
Communication Systems Engineering,
University of the Aegean,
Samos 83200
stamatatos@aegean.gr

**Pierre Zweigenbaum**
LIMSI, CNRS
Université Paris-Saclay
F-91405 Orsay
pz@limsi.fr


## Abstract


Electronic Health Records (EHRs) are increasingly available in modern health care institutions either through the direct creation of electronic documents in hospitals' health information systems, or through the digitization of historical paper records. Each EHR creation method yields the need for sophisticated text reuse detection tools in order to prepare the EHR collections for efficient secondary use relying on Natural Language Processing methods. Herein, we address the detection of two types of text reuse in French EHRs: 1) the detection of updated versions of the same document and 2) the detection of document duplicates that still bear surface differences due to OCR or de-identification processing. We present a robust text reuse detection method to automatically identify text reuse in document pairs in two French EHR corpora that achieves an overall macro F-measure of 0.68 and 0.60, respectively and correctly identifies all redundant document pairs of interest.




## 1 Introduction

Over the last decade a large number of hospitals and medical institutions have adopted the use of Electronic Health Records (EHRs) to store patient records and medical details. Simultaneously, the lowered cost of computational resources has given rise to digitization efforts of existing (paper) collections. While the presence of such large, digital corpora opens up exciting possibilities for medical data and text mining or modelization efforts, this is not without certain caveats. The resulting digital collections often are noisy, with several issues that can have an impact on the accuracy of subsequent text mining processes, such as encoding errors, missing files, OCR errors, etc. One interesting issue in cumulatively constructed text corpora is the problem of 'text reuse'. Text reuse is defined here as the intentional or unintentional reusing of existing text (fragments) to create a new text, for example, by copy-pasting text fragments from one document to fit into a new document; or by adapting a report and saving both the old and the new version as separate documents. Text reuse is a complex phenomenon which has been studied in multiple settings such as newspaper journalism (Clough et al., 2002), programming code (Ohno and Murao, 2009), the analysis of text reuse in blogs and web pages (Abdel Hamid et al., 2009), etc. It is quite prevalent in the medical domain (Wrenn et al., 2010) and often seen as a negative factor: Cohen et al. (2013) found that copy-pasting practices in US hospitals have a significant negative impact on the accuracy of the subsequent text mining systems on the clinical notes. However, when text reuse is considered as a diachronic phenomenon, it has some interesting aspects. By identifying which text (fragments) have been reused we can follow the flow of information over time in a patient's file. Moreover, adjustments that are made to copied text (fragments) can give an insight into the thought process of the acting clinicians and may help identify potential errors or adjustments during the treatment process (Hirschtick, 2006).




*Proceedings of the Fifth Workshop on Building and Evaluating Resources for Biomedical Text Mining (BioTxtM 2016),*
pages 108–114, Osaka, Japan, December 12th 2016.

Text reuse has been studied extensively in the context of authorship attribution and plagiarism detection (Stamatatos, 2009). In general we can distinguish between two main types of text reuse: 'global text reuse' in which the task is to pair up (near-) duplicate documents that exists in different locations, or whose differences are linked to version control issues; and 'local text reuse' which occurs when people borrow or plagiarize smaller text fragments such as sentences or passages from various sources to incorporate in a new text. Both types are included in this study.

While the goal is the same, there are some key differences between plagiarism detection and text reuse detection in the medical domain. Medical professionals work under an enormous time pressure, so rather than rewriting an existing text (fragment), they will merely add new information or adjust existing information, and at best edit out some orthographic errors or write out acronyms that existed in the previous version. Consequently, our methods can focus on literal string matching, rather than employing semantic similarity measures (other than detecting spelled-out variants of acronyms) or paraphrase detection. Furthermore, redundancy detection is usually performed within a closed reference collection (as opposed to plagiarism detection systems that use the entire internet as a reference base). Another difference is the quality of the written text. Depending on the quality and the nature of the text formatting tools that are available, electronic health records may contain an astounding number of orthographic errors (Ruch et al., 2003) or in the case of a digitized corpus, a large variety of OCR errors. Another source of potential minor surface variation is the de-identification process in which personal health identifiers (PHI) such as patients names, phone numbers, record numbers are replaced by plausible synthetic surrogates (Sweeney, 1996; Meystre et al., 2010). Depending on how the process is implemented, for example with the inclusion of random substitution, numbers that were the same in the original documents can appear with slight variations in the de-identified documents. Text reuse metric tools in this domain therefore need to be robust to the noise of these sources of surface variation and correctly detect similar text segments even when the surface forms do not match 100%.

The current paper presents a simple but effective tool for text reuse detection in the medical domain, both for global and local text reuse detection, which proves robust to surface variation prevalent in medical texts by allowing for character gaps when calculating the blocks of reused texts. The tool is meant to figure as a module in a larger framework, i.e. a pipeline which normalizes and extracts information from documents in a patient file in order to model the patient's treatment over time. This adds a practical component to the evaluation of the proposed tool. Missing a case of text reuse is a more grievous error than (mis)labeling a false positive. A mislabeled case, i.e either not correctly determining between different degrees of reuse, or erroneously spotting text reuse, can be spotted by the information extraction module later on in the pipeline. When a case of text reuse is not identified, however, no subsequent processing will occur for that document pair and the information is effectively lost for the information extraction process. In this paper we present and evaluate the text reuse detection tool in isolation and discuss its strengths and weaknesses.

## 2  Background

A traditional approach for the detection of verbatim copying[1] is to compute the similarity between the source and target text as the proportion of substring sequences that the two texts have in common. These substring sequences can either be defined as character n-grams (Cohen et al., 2013), words (Wrenn et al., 2010), or word n-grams (Adeel Nawab et al., 2012). These methods are mainly based on fingerprinting and hashing techniques, i.e. the documents are represented as sets of unique digital signatures, and are highly precise but are not robust to much surface variation. Some methods, however, are adapted to deal with insertions and deletion of words or characters. For example, as an extension of the 'longest common substring' algorithm (Gusfield, 1997), which calculated text similarity as the length of the longest continuous sequence of characters normalized by the sum of the document lengths, Wise et al. (1996) developed the 'Greedy String Tiling' which allows for insertions and deletions. It determines the maximum set of contiguous substrings that two documents have in common, wherein each substring has the largest possible length. However by eliminating word order through the construction of the

---

[1]As opposed to semantic reuse where the same idea of message is rewritten in a different manner.

set, valuable information on the ordering of the subsequences is lost. The method proposed in this paper aims to address this problem by allowing for a 'mismatch gap' (see section 3) while still keeping information on the original subsequence order when calculating the similarity score. Another form of surface variation that needs to be caught—especially in OCRed corpora—is due to differences in formatting: Lopresti (2000) developed a string matching algorithm that distinguishes between differences in content and differences in formatting within a document pair.

The study of text reuse detection in the medical domain has either focused on plagiarism detection in medical articles in PubMed (Errami et al., 2010; Sun et al., 2010) or for dedicated journals (Baždarić et al., 2012) or on text reuse during the creation of medical corpora and its consequences for database integrity or subsequent text mining applications (Wrenn et al., 2010). Zhang et al. (2011) found that redundant information contained in US clinical notes increases over time and that a text reuse detection tool with domain-specific knowledge is a necessary step in the detection of novel information within clinical files (Zhang et al., 2012).

## 3 Text reuse detection tool

The text reuse detection tool presented in this article consists of three main modules and is inspired by the best practices from recent research in plagiarism detection (Potthast et al., 2014). In a first step, the text is split into character n-grams of a user-defined length. Each substring unit is indexed with information on its position in the source document (character offsets). The document is thus transformed into a bag of overlapping character n-grams. We then apply a global alignment algorithm[2] to find the alignment of sequences with the largest global overlap between the two documents. In a second step, we then resolve gaps in the alignment, i.e. disjoint blocks, and construct larger blocks of aligned text. Where a large number of consecutive in-common substring sequences are detected that are interspersed by spurious non-matching blocks, the substring sequences are merged into a larger (quasi-)matching block by a user-defined 'gap parameter'. This parameter was heuristically set to 3 characters for the experiments described in this paper. For the OCRed corpus we also experimented with a variant in which larger character gaps were allowed if the non-matching blocks of the two documents contained 'confusion pairs'[3] of common OCR errors that were extracted from a training corpus. This 'gap parameter' catches small differences in formatting or character variations, i.e. a misspellings or OCR errors, between the two documents.

Finally, in the third step, the tool outputs the constructed larger 'matching blocks' with offset information for local text reuse detection, and calculates the proportion of matching text over the length of its source document to give an estimation of the global text reuse between the two documents. At this point the tool does not yet filter out text blocks that are below a certain length threshold (thus eliminating spurious matches). The use of short character n-grams (n=3) ensures that the similarity score will not be largely affected by small differences in detected fragments caused by OCR errors or differences in formatting. Figure 1 illustrates step 2 and 3 of the process.

## 4 Corpora

We show the performance of the tool on two separate and distinct corpora of French clinical notes, which exemplify the problem of local and global similarity, respectively.

### 4.1 Corpus with local text reuse (LTR)

The first corpus consists of 107 documents that describe a patient's illness, renal transplantation and follow-up case over time through various lab results, consultation reports, etc. The corpus is originally an EHR corpus, that is, the original text was edited in Word documents which were later on automatically transformed into text files using the `AntiWord`[4] tool which converts MS Word documents into plain

---

[2]Implemented in the Python difflib library

[3]Confusion pairs are systematic OCR errors in which a character or a sequence of characters in the source document is consistently replaced by another character or sequence of characters during the OCR process. For example, characters like 'i' and 'l' are visually similar and thus often confused.

[4]http://www.winfield.demon.nl/

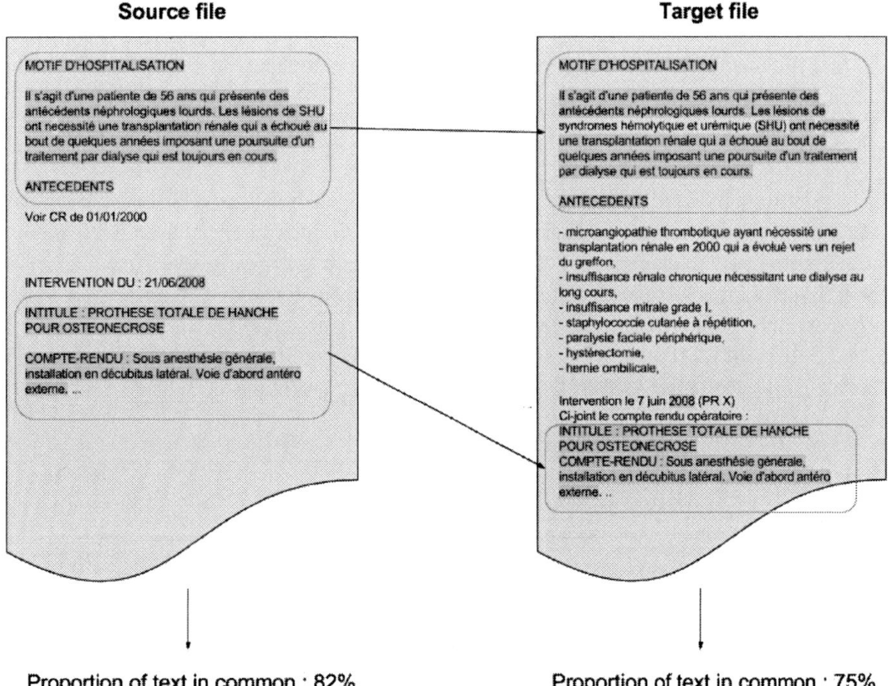



Figure 1: Report samples of the local text reuse corpus with fake data and realistic text reuse examples. The source file constitutes an earlier (i.e. older) version of the the target file on the right within the same patient records. Green highlighting indicates initial matching characters and the green blocks show the constructed 'matching blocks' with variation gaps included. For reasons of legibility we do not show the overlapping character sequences that were created in step 1. Please note that inconsistencies with regards to dates are caused by the de-identification method. The proportion of text in common that is used to calculate the similarity between the document pair is based on the entire documents, here we only show an excerpt.

text. This corpus is a subset of larger corpus which was used in a previous study on text reuse (D'hondt et al., 2015). While it does not contain a large amount of text reuse, the corpus exhibits an important temporal dimension, i.e. medical cases that span multiple years. For this reason, the corpus contains a series of documents reporting on regular check-ups that each build on the previous one, by retelling the medical history of the patient and completing it with the most recent exam results. Another interesting series of documents are follow-up exams that are conducted several times as part of the patient care pathway, and may yield similar results on each instance.

## 4.2 Corpus with global text reuse (GTR)

The second corpus consists of 1,007 documents from French foetopathology[5] reports, with data from 25 different patients. This corpus was assembled and digitized within the context of the Accordys project. The digitization effort consisted of OCRing the original typed-out pages, which was then followed by a de-identification step. There is a substantial amount of redundancy in this corpus: For some documents, several (nearly-identical[6]) copies of the same original document were added to the patient's folder. However, the de-identification process has deleted parts of the text for some copies, but not for others. While

---

[5]The medical domain which specializes in the treatment and diagnosis of illnesses in unborn children.

[6]While the original paper documents might be identical, the process of OCR and de-identification has introduced enough noise that very few identical files remain.

the patient files in the corpus do not span a long time individually, there are multiple cases in which different (intermediate) versions of a document were contained in the file. It is therefore a considerable challenge to distinguish between near-duplicate files that originate from the same original document, or those that came from two different versions of that document.

## 5  Corpus analysis

For both corpora we generated all possible document pairs for each patient. These document pairs were then labeled by two independent annotators[7] with regards to the similarity between the two documents. The annotators took care to distinguish between (near-)duplicate text[8] (category '2') and documents that are either different versions of the same report, e.g. an intermediate version versus the final version with more information, or similar reports on two different events (category '1'). Table 1 shows the labeling scheme, and the cut-off scores that were used to classify the output of the duplication detection tool. The number of document pairs for each of the three categories can be found in Tables 2 and 3 for the two different categories.

| Category label | Category description | Score cut-off |
|---|---|---|
| 2 | near-duplicates | $x >= 0.9$ |
| 1 | different version of same base document or different events | $0.5 >= x < 0.9$ |
| 0 | documents are unrelated | $x < 0.5$ |

Table 1: Explication of labels used in study.

### 5.1  Corpus with local text reuse (LTR)

| Category label | # document pairs in reference set | Precision | Recall | F1-score |
|---|---|---|---|---|
| 2 | 2 | 0.20 | 1.00 | 0.33 |
| 1 | 6 | 0.60 | 0.86 | 0.71 |
| 0 | 99 | 1.00 | 1.00 | 1.00 |
| macro-average | - | 0.60 | 0.95 | 0.68 |

Table 2: Precision and Recall scores for EHR corpus (LTR)

The local text reuse corpus only has a small number of positive examples of text reuse but the tool still categorizes the majority of the document pairs correctly. The low Precision score for category '2' is caused by the distinctive structure in the yearly follow-up reports that are included in the corpus. While the documents contain different information, i.e. one follow-up report describes the state of patient one year after the transplant, a second document describes his/her state after 5 years, they follow a similar structure and formatting and contain little free text. To correctly identify that such documents pertain to different medical events, additional information such as identifying the documents time stamps is needed. Copy-pasting of results from smaller, non-structured report into the medical overview was successfully detected however. From a medical perspective of effectively reviewing the patient record, all document pairs with some form of reuse have been successfully identified, so that the bulk of the manual review work can be lightened using this tool.

### 5.2  Corpus with global text reuse (GTR)

The second corpus contains more examples of near-duplicates. Interestingly, we find that our tool has severe problems with the detection of intermediary versions of reports, and often categorizes them as category 2 (identical pairs). A deeper analysis of the errors shows that the current method does not take

---

[7]We did not calculate IAA but few conflicting annotations occurred. Conflicts in annotations were resolved after discussion.

[8]In the case of local text reuse this can signify parts of the document, in case of global text reuse it refers to the entire document.

| Category label | # document pairs in reference set | Precision | Recall | F1-score |
|---|---|---|---|---|
| 2 | 218 | 0.96 | 0.66 | 0.78 |
| 1 | 55 | 0.04 | 0.05 | 0.05 |
| 0 | 1451 | 0.95 | 0.99 | 0.97 |
| macro-average | - | 0.65 | 0.57 | 0.60 |

Table 3: Precision and Recall scores for OCRed corpus (GTR)

document length into account. Some reports are highly similar in all sections but for the 'conclusion section'. In other document pairs, the intermediary versions are missing only one section which is present in the final version. One way of dealing with such differences between text versions would be to add a boosting factor for longer text insertion, i.e. a long block of inserted text should have a stronger (negative) impact on the similarity score than the same number of inserted characters spread out over various, shorter blocks of inserted text. This approach would certainly improve classification accuracy between the '1' and '2' categories, but the booster factor would be hard to determine with regards to the document length. A more accurate approach would be to equip the tool with additional data, either on document structure, e.g. perform the comparison on section level rather than document level, or on the time stamps of the generated documents. While many documents of category '1' are mislabeled as near-duplicates, analysis of the correct pairings in category 2 shows that the tool exhibits a high precision in extracting real near-duplicates, even in the face of a high OCR error rate (D'hondt et al., 2016).

## 6   Conclusion

In this paper we present a character-based tool for the detection of text reuse, and evaluate its usability on two different French EHR corpora. We find that our tool is robust to the surface variation in the two corpora which were introduced by OCR and orthographic errors as well as variations introduced by the de-identification process. As such we believe it is well-suited to be included in a NLP pipeline that will process a large variety of medical corpora. While the tool generally achieves a high recall score which is important for the subsequent pipeline, it lacks in precision. The tool is not able to distinguish more 'semantic' differences such as the differences between intermediate and final versions, or when reports describe highly similar yet different events. To capture such information the tool needs to be coupled with additional information in the NLP pipeline such as information on the time stamp of the document, or information on document structure, i.e. so that the tool will only be run on parts of the document that contain free text. One limitation of this study is the size of the EHR corpus used for testing the method. While the preliminary results obtained here are encouraging, they would need to be confirmed on a larger data set. We plan to address this in future work in collaboration with physicians who will also provide qualitative feedback on the usability of the tool in a clinical setting.

## Acknowledgements

This work was supported by the French National Agency for Research under the Accordys[9] ANR-12-CORD-0007, and CABeRneT[10] ANR-13-JS02-0009-01 grants.

## References

Ossama Abdel Hamid, Behshad Behzadi, Stefan Christoph, and Monika Henzinger. 2009. Detecting the origin of text segments efficiently. In *Proc of the 18th international conference on World wide web*, pages 61–70. ACM.

Rao Muhammad Adeel Nawab, Mark Stevenson, and Paul Clough. 2012. Detecting text reuse with modified and weighted n-grams. In *Proc of the First Joint Conference on Lexical and Computational Semantic (*SEM)*, pages 54–58, Montréal, QC. Association for Computational Linguistics.

---

[9]Agrégation de Contenus et de COnnaissances pour Raisonner à partir de cas dans la DYSmorphologie foetale

[10]CABeRneT: Compréhension Automatique de Textes Biomédicaux pour la Recherche Translationnelle

Ksenija Baždarić, Lidija Bilić-Zulle, Gordana Brumini, and Mladen Petrovečki. 2012. Prevalence of plagiarism in recent submissions to the Croatian Medical Journal. *Science and engineering ethics*, 18(2):223–239.

Paul Clough, Robert Gaizauskas, Scott SL Piao, and Yorick Wilks. 2002. Meter: Measuring text reuse. In *Proceedings of the 40th Annual Meeting on Association for Computational Linguistics*, pages 152–159. Association for Computational Linguistics.

Raphael Cohen, Michael Elhadad, and Noémie Elhadad. 2013. Redundancy in electronic health record corpora: analysis, impact on text mining performance and mitigation strategies. *BMC Bioinformatics*, 14:10.

Eva D'hondt, Xavier Tannier, and Aurélie Névéol. 2015. Redundancy in French electronic health records: A preliminary study. In *Proc of the 6th Health Text Mining and Information Analysis Work (LOUHI)*, pages 21–30, Lisbon, Portugal.

Eva D'hondt, Cyril Grouin, and Brigitte Grau. 2016. Redundancy in French electronic health records: A preliminary study. In *Proc of the 7th Health Text Mining and Information Analysis Work (LOUHI)*, pages 61–68, Austin, TX.

Mounir Errami, Zhaohui Sun, Angela C George, Tara C Long, Michael A Skinner, Jonathan D Wren, and Harold R Garner. 2010. Identifying duplicate content using statistically improbable phrases. *Bioinformatics*, 26(11):1453–1457.

Dan Gusfield. 1997. *Algorithms on strings, trees and sequences: computer science and computational biology*. Cambridge university press.

Robert E Hirschtick. 2006. Copy-and-paste. *Jama*, 295(20):2335–2336.

Daniel P Lopresti. 2000. String techniques for detecting duplicates in document databases. *International Journal on Document Analysis and Recognition*, 2(4):186–199.

Stephane M Meystre, F Jeffrey Friedlin, Brett R South, Shuying Shen, and Matthew H Samore. 2010. Automatic de-identification of textual documents in the electronic health record: a review of recent research. *BMC Med Res Methodol*, 10(70).

Asako Ohno and Hajime Murao. 2009. A new similarity measure for in-class source code plagiarism detection. *International Journal of Innovative Computing, Information and Control*, 5(11):4237–4247.

Martin Potthast, Matthias Hagen, Anna Beyer, Matthias Busse, Martin Tippmann, Paolo Rosso, and Benno Stein. 2014. Overview of the 6th international competition on plagiarism detection. In *Working Notes for CLEF 2014 Conference*, pages 845–876, Sheffield, UK.

Patrick Ruch, Robert Baud, and Antoine Geissbühler. 2003. Using lexical disambiguation and named-entity recognition to improve spelling correction in the electronic patient record. *Artificial intelligence in medicine*, 29(1):169–184.

Efstathios Stamatatos. 2009. A survey of modern authorship attribution methods. *Journal of the American Society for information Science and Technology*, 60(3):538–556.

Zhaohui Sun, Mounir Errami, Tara Long, Chris Renard, Nishant Choradia, and Harold Garner. 2010. Systematic characterizations of text similarity in full text biomedical publications. *PLoS One*, 5(9):e12704.

Latanya Sweeney. 1996. Replacing personally-identifying information in medical records, the Scrub system. In *Proceedings of the AMIA annual fall symposium*, pages 333–337. American Medical Informatics Association.

Michael J Wise. 1996. YAP3: Improved detection of similarities in computer program and other texts. *ACM SIGCSE Bulletin*, 28(1):130–134.

Jesse O Wrenn, Daniel M Stein, Suzanne Bakken, and Peter D Stetson. 2010. Quantifying clinical narrative redundancy in an electronic health record. *Journal of the American Medical Informatics Association*, 17(1):4953.

Rui Zhang, Serguei Pakhomov, Bridget T McInnes, and Genevieve B Melton. 2011. Evaluating measures of redundancy in clinical texts. In *AMIA Annual Symposium Proceedings*, volume 2011, page 1612. American Medical Informatics Association.

Rui Zhang, Serguei Pakhomov, and Genevieve B. Melton. 2012. Automated identification of relevant new information in clinical narrative. In *Proc of the 2nd SIGHIT International Health Informatics Symposium (IHI)*, pages 837–842, Miami, Florida, USA. ACM.

# Negation Detection in Clinical Reports Written in German

Viviana Cotik[§], Roland Roller[‡], Feiyu Xu[‡], Hans Uszkoreit[‡], Klemens Budde[◇] and Danilo Schmidt[◇]

[§]Universidad de Buenos Aires, Buenos Aires, Argentina
[§]Universität des Saarlandes, Saarbrücken, Germany
[§]Language Technology Lab, DFKI, Berlin, Germany
vcotik@dc.uba.ar
[‡]Language Technology Lab, DFKI, Berlin, Germany
{firstname.surname}@dfki.de
[◇]Charité Universitätsmedizin, Berlin, Germany
{firstname.surname}@charite.de

## Abstract

An important subtask in clinical text mining tries to identify whether a clinical finding is expressed as present, absent or unsure in a text. This work presents a system for detecting mentions of clinical findings that are negated or just speculated. The system has been applied to two different types of German clinical texts: clinical notes and discharge summaries. Our approach is built on top of NegEx, a well known algorithm for identifying non-factive mentions of medical findings. In this work, we adjust a previous adaptation of NegEx to German and evaluate the system on our data to detect negation and speculation. The results are compared to a baseline algorithm and are analyzed for both types of clinical documents. Our system achieves an F1-Score above 0.9 on both types of reports.

## 1 Introduction

Named entity recognition (NER) and relation extraction (RE) are central research topics in medical text mining. Clinical reports often contain a large number of expressions of negation and speculation. It is important to recognize whether extracted assertions (especially on medical conditions) describe these findings as factual, as contrafactual (absent) or as speculated (suspected). If, for instance, the report mentions *urolithiasis* (kidney stones) it surely matters, whether this medical condition has been diagnosed, rejected or merely suspected.

In comparison to many other text types, electronic health reports, radiology reports and other kinds of medical reports are often written in a rather telegraphic style. Furthermore they contain many technical terms as well as non-standard and ambiguous abbreviations (Kim et al., 2011). Many of those issues also appear in social media texts (Reitan et al., 2015). However, in the biomedical domain there are only very few annotated corpora available, due to data privacy issues. Therefore the curation or development of suitable data and tools for the clinical domain pose great challenges.

Various tools have been created for detecting negations and speculations in English medical reports. Probably the most popular one is NegEx (Chapman et al., 2001). The algorithm takes as input sentences with tagged *findings* and a list of negation and speculation terms called *triggers* and then determines whether the *finding* is within the scope of negation or speculation. In comparison to English, German clinical data differs in various characteristics which have to be taken into account for the successful application of an algorithm detecting non-factuality. First of all, German is a richly inflected language (e.g. *no* can be translated as *kein, keiner, keine* etc.). Furthermore, German includes *discontinuous triggers*, such as *kann ... ausgeschlossen werden ...*[1] (*can be ruled out*). Triggers may precede, but may also follow the negated expression, as presented in Table 1. Regarding this situation, Wiegand et al. (2010) state, that the detection of negation scope in German language is more difficult than in other languages, such as English.

---



[1]Dots indicate potential positions of the finding: (kann ... *finding*... ausgeschlossen werden, ... *finding*... kann ausgeschlossen werden)



*Proceedings of the Fifth Workshop on Building and Evaluating Resources for Biomedical Text Mining (BioTxtM 2016),*
pages 115–124, Osaka, Japan, December 12th 2016.

| precede | follow |
| --- | --- |
| *frei* von Beschwerden<br>(*free of symptoms*) | *beschwerdefrei*<br>(*without symptoms*) |
| ***nicht*** *klopfschmerzhaft*<br>(*no percussion tenderness*) | *Hinweise für eine cerebrale* Metastasierung *gibt es derzeit **nicht**.*<br>(*There is no indication of a cerebral metastasis.*) |

Table 1: Same negation triggers that might precede or follow a finding

Another interesting aspect of German negations are *surrounding triggers*, such as *lehnt ... ab (reject)* and *wies ... zurück (declined)).* In many cases it is possible to reduce/shorten triggers. However, in the case of given examples, a reduction would make the triggers too general, extending them to different meaning: *wies* (without *zurück*) for instance, could mean *to reject*, but also *to verify* in combination with the separated particle *nach*. Similar to English, negations can be directly bound to a target word as prefix or suffix, such as ***un**auffällig (**un**remarkable)*, fett***frei** (**non**fat) or motivations**los** (**without** motivation)*.

In this paper we present an adaptation of NegEx (Chapman et al., 2001) for German clinical notes and discharge summaries. Our work is based on a previous version of NegEx triggers translated to German (Chapman et al., 2013). We conducted the following modifications: 1) we corrected and extended the trigger set, 2) we extended the regular expressions to possible expansions, and 3) we classified the triggers according to their position relative to the findings. Our work differs from Chapman et al. (2013) in that we evaluate NegEx on German clinical texts. The evaluation is carried out on two types of clinical data sets (clinical notes and discharge summaries) and it is compared to a baseline algorithm.For evaluation purposes we created a gold standard. Our system outperforms the baseline on both document types and achieves a F1-Score of over 0.9.

The remainder of the paper is organized as follows. Section 2 presents previous work in the detection of negation terms in the medical domain. Section 3 presents the main contributions, by explaining the methods and the data sets used, by providing an analysis of length and types of negation and speculation terms and by describing the generation of our gold standard. Section 4 presents the results of evaluating each of the algorithms with the test data set. After a discussion of the obtained results, the paper ends with conclusions and an outlook on future work.

## 2   Previous Work

Negation detection in the biomedical domain is a well-studied problem. Various workshops and challenges have addressed this problem in the last years, such as the *Workshop on Negation and Speculation in Natural Language Processing in 2010*,[2] *CoNLL 2010 Shared Task: Learning to Detect Hedges and Their Scope in Natural Language Text* (Farkas et al., 2010), the *2010 i2b2 NLP challenge*, that focused on the negation and uncertainty identification (Uzuner et al., 2011) and *SEM 2012 Shared Task: Resolving the Scope and Focus of Negation* (Morante and Blanco, 2012). Díaz published a book about Negation and Speculation detection in medical texts (2014). S. M. Meystre and Hurdle (2008) present a review of information extraction in biomedical texts, which also addresses negation detection.

A widely used tool for negation and speculation detection is Negex (Chapman et al., 2001). The method uses a simple algorithm based on regular expressions to detect *triggers* that indicate negation or speculation. Next it uses a window of words preceding or following each relevant term to determine if the term is under the scope of negation or speculation or not. NegEx has been extended to Context (Harkema et al., 2009) and adapted to Swedish, French, Spanish and other languages with good results (Skeppstedt, 2011; Deléger and Grouin, 2012; Cotik et al., 2016; Stricker et al., 2015; Costumero et al., 2014; Afzal et al., 2014). Beside NegEx and Context a wide range of other methods exist, e.g. based on syntactic techniques (Huang and Lowe, 2007; Mehrabi et al., 2015; Sohn et al., 2012; Cotik et al., 2016) or machine learning techniques (Uzuner et al., 2009). However, in clinical context simple methods, such as NegEx work very reliably for the task they have been designed for.

---

[2]`http://www.clips.ua.ac.be/NeSpNLP2010/program.html`

116

Other research has been dedicated to clinical negation detection together with the detection of pathological entities in German texts. Bretschneider et al. (2013) classify sentences containing pathological and non-pathological findings in German radiology reports. Their approach uses a syntacto-semantic parsing approach. Gros and Stede (2013) present Negtopus, a system that identifies negations and their scope in medical diagnoses written in German and in English.

Chapman et al. (2013) translate NegEx triggers into Swedish, French and German. The work reports, among others, the frequency of occurrence of German triggers in an annotated corpus of German medical text (Wermter and Hahn, 2004), that, as far as we know, is not available for public use. Both publications, (Gros and Stede, 2013) and (Chapman et al., 2013), are related to our work. However, Negtopus focuses currently only on negation terms. It has been evaluated on a set of only 12 cardiology reports for German negation detection. NegEx with the German trigger set has not been evaluated and thus its performance is still unknown to us.

## 3 Methods

The adaptation of NegEx to German requires having a set of triggers written in German. In order to evaluate the new system, a gold standard data set is necessary, consisting of medical text with tagged *findings* and a classification of those *findings* as negated, speculated or affirmed.

### 3.1 Baseline Algorithm description

The baseline algorithm uses a small list of negation and speculation terms obtained from a previous annotation task of another dataset. If one of those terms co-occur in the same sentence with a previously tagged *finding*, we assume the *finding* is negated or speculated. If not we assume it is affirmed.

### 3.2 NegEx Algorithm description

NegEx (Chapman et al., 2001) takes as input sentences, each of them with a previously tagged *finding*, and a list of triggers (*negation and speculation terms*), and as output it determines whether the finding is negated, speculated or affirmed. Each trigger has a label assigned, which determines the scope of the negation or speculation. PREN and POST labels correspond to negation terms that occur before and after the finding respectively. The same occurs with PREP and POSP, referring to speculation terms. CONJ refers to trigger terms that terminate the scope of a negation or speculation and PSEU to pseudo-negations.[3] For more information refer to Chapman et al. (2001).

The algorithm takes the following decisions: if a finding appears more than once in the sentence, and one of the occurrences is negated, the algorithm assumes that all occurrences are negated. If there are many occurrences of the same trigger in the trigger list (with different labels), the algorithm uses the label according to this precedence list: PREN, POST, PREP and POSP.

### 3.3 Triggers

The translated NegEx triggers of Chapman et al. (2013) are publicly available and our work is based on them. However, due to various reasons the original translation has been adapted by us. First, in some cases the authors suggest alternative formulations and regular expressions for a trigger. Those alternatives were added to the trigger list and regular expressions were transformed into strings (e.g. *kein.*$\{0,2\}$*signifikant.*$\{0,2\}$*(aenderung.*$\{0,2\}$|*Veraenderung.*$\{0,2\}$) to *keine signifikante aendeurng | keine signifikanten anderungen*, etc.) (*no significant changes*). Next, a small set of triggers have been exchanged by using an alternative translation. Moreover, new triggers which appeared to be useful were also added to the list. Classification with respect to speculation, proper negation and pseudo-negations and direction of scope was also revised for all triggers (i.e. the appropriate labels were assigned). A set of 506 triggers was obtained.[4] In addition to our trigger set, tests were also performed with the triggers translated by Chapman et al. (2013) without modification. The set contains 167 triggers. Alternative translations and regular expressions were not considered.

---

[3] If a *finding* is under the scope of a PSEU trigger, NegEx assumes it is affirmed.

[4] The link to the trigger data set will be made available here: `http://macss.dfki.de`.

### 3.4 Creation of a German Negation and Speculation Gold Standard

The data used for the following experiments consists of anonymized German discharge summaries and clinical notes of the nephrology domain. Both types of documents (discharge summaries and clinical notes) are written by medical doctors and have significant differences. The clinical notes are rather short and are written by doctors during or shortly after a visit of a patient. Discharge summaries instead are written during a stay at the hospital. The document is more structured. It contains information about medical history, diagnosis, condition, medication etc. of the patient. Discharge summaries contain much more text compared to clinical notes and often contain longer and more well-formed sentences.

Both types of documents exhibit non-standard abbreviations, that might include findings and negations among them (e.g. *oB -ohne Befund (without finding)-, opB -ohne pathologischer Befund (without pathological finding)-*). Texts have morphemes representing negation, speculation or findings and positioned as prefix, suffix or in the middle of a word. Examples are *un**, like in *unangenehm (uncomfortable)*, *unklar(e—er—es) (not clear)*, *unverändert(e) (unchanged)* and **los* or **losigkeit*, like in *Appetitslosigkeit (anorexia)* and *Schlaflosigkeit (insomnia)* (both represent findings), *problemlos(e) (without problems)* (that represents the absence of a finding). Table 2 provides an overview of the annotated data set used to test our experiment.[5]

|  | discharge summaries | clinical notes |
|---|---|---|
| # number of documents | 8 | 175 |
| total amount of words | 6221 | 6674 |
| total amount of sentences | 1076 | 1158 |
| avg. words per document (std. deviation) | 777.63 (322.14) | 38.14 (30.49) |

Table 2: Comparison of annotated data sources

In order to be able to evaluate the results of our NegEx adaptation, a manually annotated gold standard was required. The annotation was carried out using the brat rapid annotation tool.[6] Moreover, in order to decrease the time dedicated to manual annotation, the data was automatically pre-annotated using an annotation tool (Roller et al., 2016).

Potential triggers were detected by using a small negation and speculation dictionary. Findings were pre-annotated using data of the UMLS[7] Methathesaurus. If a given string can be found in UMLS and its semantic type matches a set of predefined types (Anatomical Abnormality, Congenital Abnormality, Acquired Abnormality, Finding, Sign or Symptom, Pathologic Function, Disease or Syndrome, Mental or Behavioral Dysfunction, Neoplastic Process, Injury or Poisoning), then the string was annotated as a *finding* by the tool. After, the data was processed by a human annotator. Annotations wrongly made by the tool were removed or corrected and missing concepts were included. Furthermore, the annotator had to decide and annotate whether a given finding occurs in a positive, negative or rather speculative context. Finally, the annotations were corrected by a second -more-experienced- annotator to enhance the quality of the data.

Table 3 shows the number of findings that are affirmed, the number that are speculated and the number that are negated in the gold standard. The table shows, that both document types contain a large number of negations. It is interesting to note, that the ratio of affirmed and negated/speculated concepts is very different in both sets. While clinical notes contain approx. 25% more negations than affirmations, the data set contains hardly any speculations. On the other hand, the discharge summaries contain three times more affirmations than negations and speculations. However, the number of speculations is significantly higher compared to the clinical notes.

Table 4 and Table 5 present an analysis of the annotated negation and speculation terms for each document type. The tables depict the most frequent negation and speculation triggers in combination with

---



[5]The information was generated by applying a German tokenizer and a sentence splitter. All non alphabetical tokens were removed.

[6]http://brat.nlplab.org/

[7]https://www.nlm.nih.gov/research/umls/

| type of finding | discharge summaries | clinical notes |
| --- | --- | --- |
| affirmed | 390 | 255 |
| negated | 106 | 337 |
| speculated | 22 | 4 |
| findings (distinct) | 518 (366) | 596 (205) |

Table 3: Number of affirmed, speculated and negated findings in the gold standard.

trigger order (i.e. the trigger comes before or after the finding) and its overall frequency. Furthermore, the tables present the mean word distance between trigger and finding, including standard deviation (std) and the overall information about how frequently a trigger occurs before (b) or after (a). Table 4, for instance, shows that *kein Nachweis (no evidence)* is used in 14.15% of the cases as negation trigger before the finding. Furthermore the table shows that the mean word distance between trigger and finding in the discharge summaries is 0.92 with a standard deviation of 1.42. In 97% of the cases the trigger occurs before the finding in the discharge summaries.

| | discharge summaries | clinical notes |
| --- | --- | --- |
| | keine *(no, b, 35.85%)* | keine *(no, b, 64.47%)* |
| Trigger patterns | kein *(no, b, 15.09%)* | kein *(no, b, 27.99%)* |
| *(translation,* | kein Nachweis *(no evidence, b, 14.15%)* | keine *(no, a, 3.46%)* |
| *position, freq.)* | ohne *(without, b, 9.43%)* | kein *(no, a, 0.94%)* |
| | kein Hinweis *(no indication, b, 5.66%)* | ohne *(without, b, 0.63%)* |
| mean distance (std) | 0.92 (1.42) | 0.40 (5.62) |
| position (b/a) | 97% / 3% | 94% / 6% |

Table 4: Annotated negation terms

| | discharge summaries | clinical notes |
| --- | --- | --- |
| | Verdacht *(suspicion, b, 30%)* | ? *(?, a, 100%)* |
| Trigger patterns | fraglich *(doubtful, b, 10%)* | |
| *(translation,* | am ehesten *(likely, b, 10%)* | |
| *position, freq.)* | wahrscheinlich *(probable, b, 5%)* | |
| | wahrscheinlich *(probable, a, 5%)* | |
| mean distance (std) | 1.55 (1.64) | 0 (0) |
| position (b/a) | 80% / 20% | 0% / 100% |

Table 5: Annotated speculation terms

The tables show that the variation of triggers in the clinical notes is much smaller compared to the trigger variation in the discharge summaries. This can be explained by the telegraphic style of the clinical notes. In those reports, information is written very quickly, often while the patient is sitting next to the doctor. Due to time pressure and the internal use of the notes, verbose formulations are rare.

The analysis of the data and the development of the trigger set were performed in an independent way (annotated negation and speculation terms were not added as triggers).

## 4 Results

In this section we present the negation and speculation detection results of our NegEx adaptation (which we call *OTS* -our trigger set-) and the comparison against the original NegEx triggers provided by Chapman et al. (2013) (which we call *NTS* -NegEx trigger set-) and against our baseline. Results are presented in Table 6 and Table 7 and evaluated by using Accuracy, Precision, Recall and F1. In this case

True Positive (TP) refers to terms negated by the Gold Standard and correctly predicted by the methods. Furthermore, each table indicates the number of correctly and wrongly predicted instances.

| dataset | discharge summaries | | | clinical notes | | |
|---|---|---|---|---|---|---|
| algorithm | Baseline | NegEx | | Baseline | NegEx | |
| triggerset | – | NTS | OTS | – | NTS | OTS |
| TP | 103 | 65 | 99 | 333 | 123 | 328 |
| FP | 46 | 9 | 13 | 55 | 10 | 19 |
| TN | 366 | 403 | 399 | 204 | 249 | 240 |
| FN | 3 | 41 | 7 | 4 | 214 | 9 |
| Accuracy | 0.91 | 0.96 | 0.96 | 0.90 | 0.62 | 0.95 |
| Precision | 0.69 | 0.88 | 0.88 | 0.86 | 0.92 | 0.95 |
| Recall | 0.97 | 0.61 | 0.93 | 0.99 | 0.36 | 0.97 |
| F1 | 0.81 | 0.72 | 0.91 | 0.92 | 0.52 | 0.96 |

Table 6: Performance on the negation detection task for both datasets with NegEx and with the baseline. TP refers to True Positive results, FP to False Positive, TN to True Negatives and to False Negatives. NTS refers to NegEx original triggers and OTS to our trigger set.

| dataset | discharge summaries | | | clinical notes | | |
|---|---|---|---|---|---|---|
| algorithm | Baseline | NegEx | | Baseline | NegEx | |
| triggerset | – | NTS | OTS | – | NTS | OTS |
| TP | 9 | 0 | 11 | 1 | 0 | 2 |
| FP | 14 | 0 | 7 | 5 | 5 | 8 |
| TN | 482 | 496 | 489 | 587 | 587 | 584 |
| FN | 13 | 22 | 11 | 3 | 4 | 2 |
| Accuracy | 0.95 | 0.96 | 0.97 | 0.99 | 0.98 | 0.98 |
| Precision | 0.39 | 0 | 0.61 | 0.17 | 0 | 0.2 |
| Recall | 0.41 | 0 | 0.5 | 0.25 | 0 | 0.5 |
| F1 | 0.4 | 0 | 0.55 | 0.2 | 0 | 0.29 |

Table 7: Performance on the speculation detection task for both datasets with NegEx and with the baseline.

Table 8 shows the negation and speculation triggers that appear more than four times, taking into account discharge summaries and clinical notes.

## 5 Discussion

The results show, that the baseline algorithm provides promising results for the negation detection task. This might have to do with the fact that in German many of the triggers can be used before or after the finding (see Table 1). However, the results show, that in all cases the NegEx adaptation achieves better results compared to the baseline algorithm. In particular the negation and speculation detection applied to the discharge summaries leads to much better results than using the baseline algorithm. This can be explained by the fact that the discharge summaries include a larger variety of triggers, which are not covered by the baseline, but covered by the German trigger set. Moreover, discharge summaries have longer and more complex sentences, that include *CONJ* triggers, which end the scope of negation. However, the results show, that both algorithms achieve better results using the clinical notes. We believe the reason is related to the fact that clinical notes have much shorter and simpler sentences than the ones of discharge summaries. The test with the original German trigger set achieves lower results than our NegEx adaptation and our baseline. The results improve and are similar to ours (F1=0.92 for discharge summaries and 0.94 for clinical notes) if the trigger *keine* is added to NTS.

| trigger type | trigger | translation | number of occurences |
| --- | --- | --- | --- |
| negation | keine, kein | no | 471, 226 |
|  | ohne | without | 49 |
|  | nicht | not | 50 |
|  | noch | still/yet | 40 |
|  | aber | but | 18 |
|  | jedoch | but/however | 15 |
|  | bis auf | except for | 11 |
|  | entfernt | removed | 7 |
| speculation | verdacht | suspicion | 13 |
|  | ehesten, eher | rather | 13,8 |
|  | nicht sicher | not sure | 5 |
|  | ? | ? | 14 |

Table 8: Negation and speculation triggers used more than four times. Both kind of reports are taken into account.

Considering the 506 triggers of our data, only 27 occur in the clinical reports (see the ones used more than four times in Table 8). This makes us infer that the translation effort could be avoided in further adaptation of NegEx to other languages. Other works arrived to similar conclusions (Cotik et al., 2016).

Reviewing the errors, we found that syntactic analysis could improve our results. For instance, in *kein starker Krampf (no strong cramp)*, *Krampf* is under the scope of *kein (no)*, a *PREN* trigger, but *no* is actually addressing to *strong* and not to *cramp*. The use of Part of Speech tagging or dependency parsing information could help us avoid this error. Moreover, the original NegEx speculation triggers did not help us to find speculation. In fact with those triggers no speculation terms have been detected (see Table 7). Thus, a number of speculation triggers have been added to OTS. Triggers were taken from general German knowledge and from the transformation of some of the original negation triggers to their corresponding speculation triggers (e.g. *Ohne Verdacht* -without suspicion- originated *Verdacht -suspicion-*). In particular, we added the trigger *?* as a speculation term occurring after the finding, since we knew it is frequently used in the clinical notes to express uncertainty. Some False Negative results were generated by the abundance of acronyms, some of them indicating negation of findings (e.g. in oB *-ohne Befund, without finding-*, B *-Befund, finding-* was annotated as negated, but we don't have *o-ohne, whithout-* as a trigger). In all cases negation detection achieves better results than speculation detection. This might be due to the fact that there is much greater variety of triggers for indicating speculation than triggers for indicating negation. Additionally, we detected some missing triggers. In some cases two classifications of the triggers (e.g *nicht*) were possible (see Table 1). For those triggers we missed some correct classifications, where the trigger appeared in the less frequent order (for example *Lymphozele nicht mehr sichtbar , Lymphocele not visible anymore*) was classified as positive, since *nicht* was in the trigger list as a PREN trigger. See also trigger preference list in Section 3.2.

Parenthesis and commas were not included as CONJ triggers in our trigger set. After evaluating FP and FN results (see Tables 6 and 7) tests were performed including them. Including parenthesis and commas as triggers reduces the number of false positives. Consider for example those cases that use the trigger *nicht*: *Hat Nitrendipin nicht vertragen (**Flush**) (Did not tolerate Nitrendipin (flush))*. *Befinden seit Entlassung nicht gebessert, hat weiterhin **Diarrhoe*** (Condition has not been improved since discharge, has still diarrhoea). In the previous examples the findings *Flush* and *Diarrhoe* are out of the scope of negation and therefore misclassified. We also could avoid false negatives in speculation detection in cases such as *keine Oedeme (...) (serom?) (no edema (...) (serum?))*, because with our trigger set *serum?* is under the scope of *kein*. In a subsequent test, we included parenthesis and commas as CONJ triggers, which increased F1 of clinical notes to 0.98 and F1 of discharge summaries to 0.94 for negations and F1 of clinical notes to 0.62 (with a recall of 1) and F1 of discharge summaries to 0.58 for speculation.

As explained above, clinical notes are much shorter than discharge summaries. The language is less verbose, often just consisting of sequences of noun phrases with some embedded prepositional phrases. Discharge summaries in contrast contain more verbs and full sentences. Thus it is not surprising when our analysis of triggers shows that the term *kein(e) -no-* as a negative determiner is much more often used in clinical notes (571 vs. 128) whereas the sentence negation nicht (not) occurs more often in discharge summaries (32 vs 18).

Our NegEx adaptation for negations yields very good results. Although not easily comparable (because of being applied to different languages and types of medical reports), they are better than the ones obtained by the original algorithm for English clinical texts and to the adaptations done to Swedish and Spanish (in this last case only for clinical notes, discharge summaries results are similar to results obtained for Spanish). They also outperform results obtained on 12 German cardiology reports by Gros and Stede (2013). We believe that the fact of having short sentences with simple syntactic structures helps us to get good results. It should also be considered that our data set is highly redundant (some negations or negation types occur frequently). In order to improve results an hybrid method combining syntactic analysis could be used.

## 6    Conclusions

This paper presented negation and speculation detection of medical findings reported in German clinical data. Two approaches were introduced: A dictionary look-up algorithm, that was taken as a baseline and an approach based on a revised version of an existing German NegEx trigger set. Tests were also performed with triggers that were previously translated to German. The system has been tested on two different data sets, German discharge summaries and German clinical notes. In both cases the German NegEx system outperforms the baseline and achieves an F1-Score above 0.9. Furthermore this work presented an analysis of negations and speculations existing in both document types. The analysis shows, that physicians tend to use a structurally simple and precise language. Therefore the degree of lexical variation in expressing negation is very low. However, applying NegEx to other text types might turn out to be more challenging.

As Chapman et al. (2013) state, the translation of triggers to another languages has faced a number of issues. German is a language with agglutinative features, where a morpheme representing negation can be added to a word. NegEx does not address this fact. German is an inflected language, so a single term can be translated to many others, because of gender and number agreement. This increased the size of our trigger set.

One of the challenges of working with medical language is the need for careful anonymization. Texts also exhibit large numbers of technical terms and non-standardized and ambiguous abbreviations. All of this raises the efforts needed for corpus curation and annotation raising the demand for gold-standard data that can be shared.

## 7    Future Work

We plan to detect negation that is represented by bound morphemes (prefix or suffix) of relevant content words. If a lexeme *lf* stands for a medical finding according to the UMLS thesaurus, *lf+"los"* (*without*) should be considered as a negation of the finding, e.g., *schlaflos* (without sleeping), but also *lf+"los"* or *lf+"losigkeit"* could be included in the thesaurus (e.g. *Appetitslosigkeit (anorexia)* and *Schlaflosigkeit (insomnia)*), and in this case the presence of suffix or infix *los* does note indicate the absence of a finding.

We also intend to investigate the benefits of employing syntactic analyses to improve the results. Especially for the clinical notes, chunk parsing technology will have to be adapted in order to cope with the nature of this text sort.

## Acknowledgements

This research was partially supported by the German Federal Ministry of Economics and Energy (BMWi) through the project MACSS (01MD16011F) and by the Saarland University through the B5 - SFB 1102 'Information Density and Linguistic Encoding' project.

## References

Zubair Afzal, Ewoud Pons, Ning Kang, Miriam Sturkenboom, Martijn Schuemie, and Jan Kors. 2014. ContextD: an algorithm to identify contextual properties of medical terms in a Dutch clinical corpus. *BMC Bioinformatics*, 15(373):1–12.

Claudia Bretschneider, Sonja Zillner, and Matthias Hammon. 2013. Identifying pathological findings in German radiology reports using a syntacto-semantic parsing approach. In *Proc of the 2013 Workshop on Biomedical Natural Language Processing (BioNLP 2013)*, pages 27–35.

Wendy W. Chapman, Will Bridewell, Paul Hanbury, Gregory F. Cooper, and Bruce G. Buchanan. 2001. A Simple Algorithm for Identifying Negated Findings and Diseases in Discharge Summaries. *Journal of Biomedical Informatics*, 34(5):301–310.

Wendy W. Chapman, Dieter Hillert, Sumithra Velupillai, Maria Kvist, Maria Skeppstedt, Brian E. Chapman, Mike Conway, Melissa Tharp, Danielle L. Mowery, and Louise Deléger. 2013. Extending the NegEx Lexicon for Multiple Languages. In *Proceedings of the 14th World Congress on Medical and Health Informatics*, pages 677–681, Copenhagen, Denmark.

Roberto Costumero, Federico López, Consuelo Gonzalo-Martín, Marta Millan, and Ernestina Menasalvas. 2014. An Approach to Detect Negation on Medical Documents in Spanish. In *Brain Informatics and Health*, volume 8609, pages 366–375.

Viviana Cotik, Vanesa Stricker, Jorge Vivaldi, and Horacio Rodriguez. 2016. Syntactic methods for negation detection in radiology reports in Spanish. In *Proceedings of the 15th Workshop on Biomedical Natural Language Processing (BIONLP 16)*, Berlin, Germany. Association for Computational Linguistics.

Louise Deléger and Cyril Grouin. 2012. Detecting negation of medical problems in French clinical notes. In *Proceedings of the 2nd ACM SIGHIT International Health Informatics Symposium*, pages 697–702.

Noa P C Diaz. 2014. Negation and speculation detection in medical and review texts. In *SPLN*, volume 13.

Richárd Farkas, Veronika Vincze, György Móra, János Csirik, and György Szarvas. 2010. The CoNLL-2010 Shared Task: Learning to Detect Hedges and Their Scope in Natural Language Text. In *Proceedings of the Fourteenth Conference on Computational Natural Language Learning*, pages 1–12, Uppsala, Sweden.

Oliver Gros and Manfred Stede. 2013. Determining Negation Scope in German and English Medical Diagnoses. In *Nonveridicality and Evaluation*, pages 113–126.

Henk Harkema, John N. Dowling, Tyler Thornblade, and Wendy W. Chapman. 2009. ConText: An Algorithm for Determining Negation, Experiencer, and Temporal Status from Clinical Reports. *Journal of biomedical informatics*, 42(5):839–851.

Yang Huang and Henry J Lowe. 2007. A Novel Hybrid Approach to Automated Negation Detection in Clinical Radiology Reports. *Journal of the American Medical Informatics Association*, 14(3):304–311.

Youngjun Kim, John Hurdle, and Stphane M Meystre. 2011. Using UMLS lexical resources to disambiguate abbreviations in clinical text. *Annual Symposium proceedings*, 2011:715–722.

Saeed Mehrabi, Anand Krishnan, Sunghwan Sohn, Alexandra M Roch, Heidi Schmidt, Joe Kesterson, Chris Beesley, Paul Dexter, C Max Schmidt, Hongfang Liu, et al. 2015. DEEPEN: A negation detection system for clinical text incorporating dependency relation into NegEx. *Journal of biomedical informatics*, 54:213–219.

Roser Morante and Eduardo Blanco. 2012. Sem 2012 shared task: Resolving the scope and focus of negation. In *Proceedings of the First Joint Conference on Lexical and Computational Semantics - Volume 1: Proceedings of the Main Conference and the Shared Task, and Volume 2: Proceedings of the Sixth International Workshop on Semantic Evaluation*, SemEval '12, pages 265–274, Stroudsburg, PA, USA. Association for Computational Linguistics.

Johan Reitan, Jørgen Faret, Björn Gambäck, and Lars Bungum. 2015. Negation scope detection for twitter sentiment analysis. In *Proceedings of the 6th Workshop on Computational Approaches to Subjectivity, Sentiment and Social Media Analysis*, pages 99–108, Lisboa, Portugal. Association for Computational Linguistics.

Roland Roller, Hans Uszkoreit, Feiyu Xu, Laura Seiffe, Michael Mikhailov, Oliver Staek, Klemens Budde, Fabian Halleck, and Danilo Schmidt. 2016. A fine-grained corpus annotation schema of german nephrology records. In *Proceedings of the Clinical Natural Language Processing Workshop*, Osaka, Japan, December. Association for Computational Linguistics.

K. C. Kipper-Schuler S. M. Meystre, G. K. Savova and J. F. Hurdle. 2008. Extracting information from textual documents in the electronic health record: A review of recent research.

Maria Skeppstedt. 2011. Negation Detection in Swedish Clinical Text: An Adaption of NegEx to Swedish. *Journal of Biomedical Semantics*, 2(3):1–12.

Sunghwan Sohn, Stephen Wu, and Christopher G Chute. 2012. Dependency parser-based negation detection in clinical narratives. *AMIA Summits on Translational Science proceedings AMIA Summit on Translational Science*, 2012:1–8.

Vanesa Stricker, Ignacio Iacobacci, and Viviana Cotik. 2015. Negated Findings Detection in Radiology Reports in Spanish: an Adaptation of NegEx to Spanish. In *IJCAI - Workshop on Replicability and Reproducibility in Natural Language Processing: adaptative methods, resources and software*, Buenos Aires, Argentina.

Özlem Uzuner, Xiaoran Zhang, and Tawanda Sibanda. 2009. Machine Learning and Rule-based Approaches to Assertion Classification. *Journal of the American Medical Informatics Association : JAMIA*, 16(1):109–115.

Özlem Uzuner, Brett R. South, Shuying Shen, and Scott L. DuVall. 2011. 2010 i2b2/VA Challenge on Concepts, Assertions, and Relations in Clinical Text. *Journal of the American Medical Informatics Association : JAMIA*, 18(5):552–556.

Joachim Wermter and Udo Hahn. 2004. An Annotated German-Language Medical Text Corpus as Language Resource. In *Proc 4th Intl LREC Conf*, pages 473–476.

Michael Wiegand, Alexandra Balahur, Benjamin Roth, Dietrich Klakow, and Andrés Montoyo. 2010. A survey on the role of negation in sentiment analysis. In *Proceedings of the Workshop on Negation and Speculation in Natural Language Processing*, NeSp-NLP'10, pages 60–68, Stroudsburg, PA, USA. Association for Computational Linguistics.

# Scoring Disease-Medication Associations using Advanced NLP, Machine Learning, and Multiple Content Sources


**Bharath Dandala**
IBM Research

**Murthy Devarakonda**
IBM Research

**Mihaela Bornea**
IBM Research

**Christopher Nielson**
US Dept. of Veterans Affairs



## Abstract

Effective knowledge resources are critical for developing successful clinical decision support systems that alleviate the cognitive load on physicians in patient care. In this paper, we describe two new methods for building a knowledge resource of disease to medication associations. These methods use fundamentally different content and are based on advanced natural language processing and machine learning techniques. One method uses distributional semantics on large medical text, and the other uses data mining on a large number of patient records. The methods are evaluated using 25,379 unique disease-medication pairs extracted from 100 de-identified longitudinal patient records of a large multi-provider hospital system. We measured recall (R), precision (P), and F scores for positive and negative association prediction, along with coverage and accuracy. While individual methods performed well, a combined stacked classifier achieved the best performance, indicating the limitations and unique value of each resource and method. In predicting positive associations, the stacked combination significantly outperformed the baseline (a distant semi-supervised method on large medical text), achieving F scores of 0.75 versus 0.55 on the pairs seen in the patient records, and F scores of 0.69 and 0.35 on unique pairs.


## 1  Introduction

Electronic Health Record (EHR) systems have become invaluable repositories of patient information, but their poor design and inadequate functionality make it difficult for physicians to assimilate the vast amounts of data, reducing physician productivity and negatively impacting patient care [1] [2]. Advanced clinical decision support applications can reduce the cognitive load on physicians and improve patient care. These applications need medical knowledge for effective reasoning. One such knowledge is relationships (or more abstractly, associations) between diseases and medications. While Unified Medical Language System (UMLS) [3] semantic network contains manually curated entity relationships, it falls short in a few ways: its coverage is inadequate, the relations are binary, and it is not always clear how far to traverse in the network. An automated association scoring method that provides high coverage and accuracy is highly desirable and can be used to build a useful knowledge resource.

There are many uses for such a method and knowledge resource in clinical applications because these associations are not explicitly maintained in a typical patient record. The method (or the resource) can be used in a patient record summary to show clinicians which medications are related to a patient's active medical conditions. It can also be used in developing cohort models and for predicting disease likelihood and progression using probabilistic graphical models. In this paper, we present two new methods for scoring associations between diseases and medications, and assess their accuracy and coverage.

One of the methods is based on mining ordered medications data in millions of patient records and leveraging the temporality of events, such as disease diagnosis and medication ordering. The data mining produces statistical measures for disease and medication pairs, which are then used as features in a supervised machine learning algorithm for association scoring between a disease and a medication. A learned F1-optimized threshold is then used to classify positive and negative associations.

The second method is based on features obtained with distributional semantics on a large medical text, complemented with features from UMLS. Distributional semantics is used to obtain synonyms,






*Proceedings of the Fifth Workshop on Building and Evaluating Resources for Biomedical Text Mining (BioTxtM 2016),*
pages 125–133, Osaka, Japan, December 12th 2016.

relations between words, and to develop a taxonomy of the concepts in the domain. From UMLS, semantic types and relations of the entities are retrieved as features. Once again, a supervised machine learning model provides a score for the association, and a threshold is also learned to classify positive and negative associations.

Two aspects of relations need to be considered – type and context. Given two entities, the type of relationship between them can be specific (such as "treats", "prevents", or "causes") or it can be anonymous. Further, a relationship between two given entities can be contextual in that a specific passage may entail a specific relationship between the entities, but this may or may not hold true in a larger corpus. This paper concerns itself with *anonymous* and *context independent* relationships, which we call associations.

We conducted accuracy and coverage analysis using entity pairs from 100 de-identified patient records provided to us by a large multidisciplinary hospital system. The diseases and medications of each patient were paired and these pairs formed the data set for this study. Medical experts manually labeled each unique pair in the data set. We conducted 10 x 10 cross validations, calculated standard precision, recall, F1 scores, and coverage measures, and plotted P-R curves.

Results showed that the distributional semantics method provided higher recall, the data mining approach provided higher precision and the stacked ensemble of the two methods achieved the overall best performance. Both methods outperformed a previously reported baseline method that uses manifold models and distance learning on a large medical corpus.

## 2    Related Work

In general, entity associations can be found in human readable form in many sources. Medical textbooks, journal papers, and web content include discourse that states or implies associations. Formal documents, such as the FDA drug labels, are more organized textual resources. Patient records themselves are another valuable source. In addition, UMLS contains relationships such as "treats" and "diagnostic-of". There is a need for automated method(s) to leverage these sources.

Many automated methods exist for relation extraction from passages in general text; a recent review [4] summarizes the research. The 2010 i2b2/VA challenge [5] included extraction of a specific set of relations from clinical notes. However, more work is needed to create a knowledge method or resource for clinical applications. One recent study [6], which we use as the baseline, successfully used manifold models and distance learning to extract seven frequent relations (defined in UMLS) from medical text with the intention of creating a knowledge resource, however, its coverage was limited.

Another relevant system and method is MEDI [7], which builds an indication to prescribable medications association resource using four public resources - RxNorm, Side Effect Resource (SIDER) 2, MedlinePlus, and Wikipedia. The resources are treated as separate voting entities in this approach, which led to the conclusion that either the highest accuracy can be achieved with limited coverage (when all resources contain the entities) or that moderate accuracy can be achieved with a higher coverage (when only fewer resources contain the entities). In contrast, the methods described here automatically "learn" optimal use of the underlying resources.

In [8] [9], association rule mining from EHR records was used to extract medication to disease relationships. The fundamental strategy in these studies was to use co-occurrence of medication orders and patient problems as a source to automatically build association rules between medications and problems. In one of the two methods studied here, such a co-occurrence of medications and problems in a patient record is extracted as one of several mined statistics. A systematic review of existing medication to indication (i.e. a symptom or a diagnosis) knowledge bases and their appraisal was presented in [10]. While an abstract appraisal is useful, here, we attempt a quantitative accuracy analysis with clinical decision support applications in mind.

## 3    Methods and Experiments

### 3.1    Distributional Relation Extraction (DRE) Method

Distributional Relation Extraction (DRE) is a supervised machine learning method for discovering associations between given entity pairs using distributional semantics and UMLS. Some of its features are derived from distributional semantics applied to a large medical corpus, and the remaining features are derived from UMLS. Let us assume that DRE is attempting to determine the strength of the association between two arguments, disease *Hyperlipidemia* and medication *Simvastatin*. Figure 1 shows the features generated for the two arguments. Note that the feature space is sparse and high dimensional. The feature space is described below:

**UMLS Type Features**. The intuition is that the types of the arguments (in a taxonomy) are important constraints for association scoring, as most of the relations hold between the entities of specific types. For example, given the relation *may_treat* between two arguments, it is expected that the type of the first argument is a Medication, Chemical, Drug, etc. and the type of the second argument is a Disease, Syndrome, or Disorder. UMLS taxonomies are used to obtain one set of argument type features (the second set is described below). Since types in UMLS have multiple levels of granularity, DRE uses multi-granular features: semantic groups for coarse granularity, semantic types for medium granularity and MeSH (Medical Subject Heading) types for fine granularity. The type features are binary valued and so they have a value of 1 when present. In Figure 1, notice that for the first argument, *Simvastatin*, T1-C0003277 (with label *Cholesterol Inhibitors*) is the MSH type, ST1-T121 (with label *Pharmacological Substance*) is the semantic type and SG1-CHEM (with label *Chemicals and Drugs*) is the semantic group. We experimented using combinations of UMLS types for the arguments, but did not see significant performance improvement.

**Distributional Semantics (DS) Type Features**. DRE features also include types induced by distributional semantics using the text corpora for the arguments, and the distributional semantics tool used here is called JoBimText [11], which is an open source project. The JoBimText tool provides a framework for creating a distributional semantics resource from large corpora, from which we obtain relations between words, similar terms or pseudo-synonyms for a word, and a taxonomy for the domain. We built the JoBimText resource as described in [12] by preprocessing the text corpora available for our project. JoBimText uses a dependency parser adapted for the medical domain [13] for identifying syntactic relations, and the baseline relation extraction system mentioned earlier [6]. The JoBimText framework provides an API to access the resource built in this way. Unlike the UMLS types, there is only a single level of granularity for the DS types. But, each term may have multiple types. The DS type features are determined for both arguments, as shown in Figure 1. For Simvastatin, the DS type features are T1-Medication, T1-Treatment and T1-Inhibitor.

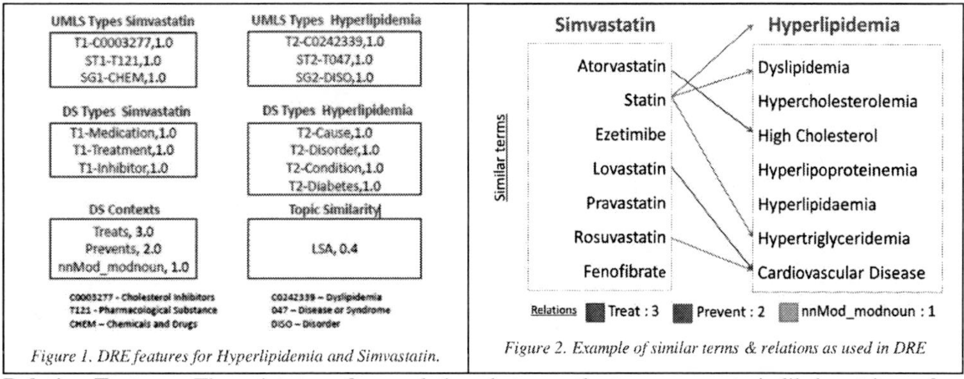


*Figure 1. DRE features for Hyperlipidemia and Simvastatin.*

*Figure 2. Example of similar terms & relations as used in DRE*

**Relation Features**. The existence of any relations between the two arguments is likely a trigger for other relations. For example, knowing that a drug prevents a disease is an indication for a "treats" relation as well. As mentioned above, the JoBimText tool can be used to identify relations that exist between the arguments in the corpus. However, a specific pair of arguments may not be mentioned together in the corpus (as processed by JoBimText) often enough, but pairs of similar terms may be present. When such expanded term pairs are considered, the relation features may provide a stronger signal. Therefore, in DRE, each relation argument is first expanded to its similar terms. As shown in Figure 2, *Simvastatin* is expanded to seven other medications including *Atorvastatin* and *Statin*, and *Hyperlipidemia* is expanded to seven other diseases including *Dyslipidemia* and *Hypercholesterolemia*. The similar term expansion is done by using the JoBimText tool. The number of expanded terms are limited to the top 10 relevant terms, based on empirical observation of optimal precision-recall trade-off in our study. Among the expanded term pairs, DRE finds three may_treat relations, two may_prevent relations, and one *nnMod_modnoun* relation (a syntactic relation obtained by analyzing the parse tree) using JoBimText as shown in Figure 2. The counts for these three relations become feature values for the relation features as shown in Figure 1.

**Topic Similarity Features**. The topic similarity between arguments can be a useful feature to detect semantic relations between them [14]. Topic similarity does not explain why things are related but does provide an indication of the presence of some relation between them. For example, Cholesterol and Diabetes are related, but their topic similarity is the same regardless of whether the relation is Diagnose or Treat. We scored the topic similarity between the two arguments using Latent Semantic Analysis (LSA) [15] [16]. The value of the LSA feature is equal to the LSA similarity between the two arguments. For the *Simvastatin-Hyperlipidemia* example, the value of the LSA similarity is 0.4.

### 3.2 Association Data (AD) method

The second association scoring method is based on the intuition that the historical, actual patient care data for a medical problem indicates clinically relevant associations between patients' problems and medications/drug-classes. This method uses a set of statistical measures obtained by mining structured and coded data in approximately six million, longitudinal patient records as features in a supervised machine learning model. The features are described below.

(In the equations below, the following notations are used:

$X_D\,/\,\overline{X}_D$ = patient records with/without an order for X and diagnosis of disease D,

$X_{\overline{D}}\,/\,\overline{X}_{\overline{D}}$ = patient records with/without an order for X and diagnosis of a disease other than D,

subscript A_D means after diagnosis D, subscript B_D means before diagnosis D,

and the time window for "at", "before", and "after" are specified in the feature definitions.)

**Frequency at diagnosis**. The fraction of patients who received an order for the given medication or its drug class among the patients who have been diagnosed with a given disease. The time window for the order is three months before or two days after.

$$FreqAtDx(D,X) = \frac{X_D}{X_D + \overline{X}_D}$$

**Relative Frequency at diagnosis**. The fraction of patients who received an order for the medication or its drug class among the patients who have been diagnosed with a given disease relative to the other diseases. The time window is the same as above.

$$RelFreqAtDx(D,X) = \frac{X_D}{X_{\overline{D}}}$$

**After versus Before diagnosis**. The ratio of the number of times the medication or its drug class was ordered before the diagnosis of a given disease to the number of times the treatment or test was ordered after the diagnosis. If the after count is zero, then the ratio is set to the maximum value observed. The time window is three months for "after" and three months for "before".

$$AfterVsBeforeDx(D,X) = \frac{X_{A\_D}}{X_{B\_D}}$$

**Odds Ratio at diagnosis**. The odds ratio of receiving the medication or its drug class at diagnosis relative to the diagnosis of the given disease. The time window for "at" diagnosis is 30 days before and two days after.

$$OddsRtAtDx(D,X) = \frac{X_D/\overline{X}_D}{X_{\overline{D}}/\overline{X}_{\overline{D}}}$$

**Odds Ratio Before diagnosis**. The odds ratio of receiving the given medication or its drug class relative to before/at the diagnosis of the given disease. The time window is three months for "before" diagnosis and 30 days before and two days after for "at" diagnosis.

$$OddsRtBeforeDx(D,X) = \frac{X_{B\_D}/\overline{X}_{B\_D}}{X_D/\overline{X}_D}$$

**Odds Ratio After versus Before diagnosis**. The odds ratio of receiving the given medication or its drug class relative to before/after diagnosis of the given disease. The time window is three months before for the "before" diagnosis and three months after for the "after" diagnosis.

$$OddsRtAfterVsBeforeDx = \frac{X_{A\_D}/\overline{X}_{A\_D}}{X_{B\_D}/\overline{X}_{B\_D}}$$

**Number of Patients Ordered.** Total number of patients who received an order for the medication or its drug class within three months before to three months after the first diagnosis of the disease.

$$N(D,X) = X_D$$

**Pearson Product-Moment Correlation.** This feature is the Pearson correlation value between the given disease (D) diagnosis and ordering the given medication or drug class (X) at the time of diagnosis. The Pearson product-moment correlation is calculated using the standard formula, using diagnoses data set $\{d_i\}, i = 1..m$, where $d_i$ is 1 if D is diagnosed in the patient record $i$ and 0 otherwise, and for each orders data set $\{x_i\}, i = 1..m$ where $x_i$ is 1 if X is ordered in the corresponding patient record $i$ and 0 otherwise. The number of patient records in the data set is $m$ for both data sets.

**Jaccard Index.** This feature is the Jaccard index calculated between the diagnoses set and the medications set for a given disease (D) and medication (X).

$$JaccardIndex(D,X) = \frac{\{d_i\} \cap \{x_i\}}{\{d_i\} \cup \{x_i\}} = \frac{X_D}{(X_D + X_{\overline{D}} + \overline{X}_D)}$$

**Arguments Expansion and Feature Aggregation**

Since the arguments to relation scoring are words and phrases (terms), it is necessary to map them to a standardized form so that the arguments can be matched with the data in the patient records. Therefore, all terms are first linked to one or more UMLS concept unique identifiers (CUIs) and then diseases are mapped to ICD9 codes and medications to RxNORM. In most cases, these mappings are one to many. So, for a given pair of disease and medication terms, $(D, M)$, the method first generates standardized pairs $\{(D_i, M_j), i = 1..n, j = 1..m\}$, and then the above defined statistical measures for each pair of standardized entities. The next step is to aggregate these n x m feature vectors into a single vector for the $(D, M)$ using the decaying sum (where $decay(p_0 \dots p_g) = \sum_{i=0}^{g} \frac{p_i}{2^i}$) which produces a single vector, $\{a_1, \dots, a_k\}$. This process is shown in Figure 3.

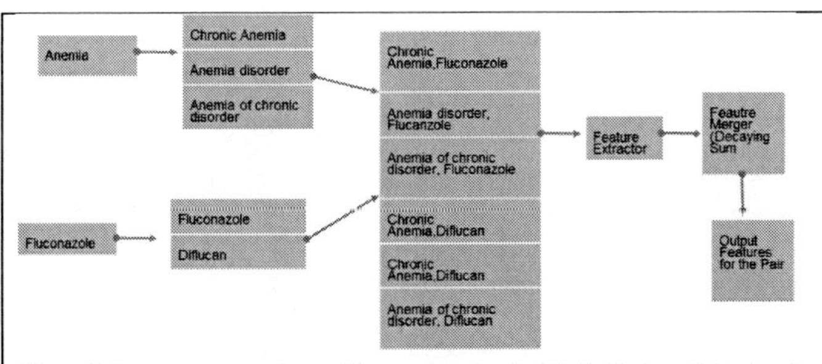


*Figure 3. Argument expansion and feature merging in AD; In Feature Extractor, for each expanded pair, the set of features described in Section 3.2 are retrieved from the association data which was previously mined from patient records; In Feature Merger, the feature values of all expanded pairs are aggregated using the decaying sum (see the text) method to create a single feature vector for the given arguments.*

### 3.3 Ensemble Method

A stacked ensemble of DRE and AD methods, designated as DRE+AD, was created. Individual scores are used as features, and a supervised machine learning model learned the optimal way to combine these scores and hence the methods and their sources. This approach is known as stacking. We used stacking (rather than combining features from both sources in a single model) for two reasons: (1) DRE and AD are fundamentally two different approaches and we wanted to study them separately and in combination; (2) DRE and AD achieved optimum performance with different machine learning methods – AD performed well with Random Forest because it uses a small number of features whereas DRE uses a high dimensional feature set which makes Logistic Regression a more suitable and effective approach.

### 3.4 Models and Training

For AD, a random forest [17] model was built since it provided the best accuracy. For DRE, DRE+AD, and for threshold learning, logistic regression models were accurate and were built with the LIBLINEAR package [18]. The training data set contained positive and negative examples of entity pairs labeled as associated (positive) or not associated (negative). As discussed below, positive and negative associations were imbalanced in the ground truth. We randomly sampled the larger set to create balanced training data.

## 3.5 Data Preparation

For this assessment, approximately 122,374 disease-medication pairs, were extracted automatically from 100 de-identified, actual patient records that were obtained under an IRB approval from a large, multi-specialty hospital. Data characteristics are summarized in Table 1. The diseases for each patient record are obtained using an automated problem list generation [19] [20] but can be replaced by the diagnostic codes or other means. The medications are taken from the medication orders in each patient record and are represented as text strings and RxNORM codes (as entered in the medication order). Therefore, if a patient record has D diseases in the problem list and M medications in the medication orders then the patient record yields $D \times M$ disease-medication pairs.

Two characteristics of this data are worth noting: (1) It contains duplicate pairs; (2) Negative examples (entities that don't have an association) are significantly larger than the positive examples. However, the data set is representative of the association scoring system input in realistic clinical applications. To remedy the duplication and asymmetry, we present results first without duplicates and later show the impact of the occurrence frequency on accuracy. Furthermore, we separately report accuracy for positively associated pairs and negatively associated pairs in the gold standard.

*Table 1. Description of the data used in this study.*

| Description | Value |
| --- | --- |
| Patient records | 100 |
| Disease-medication pairs from the patient records | 122,374 |
| De-duplicated disease-medication pairs | 25,379 |
| Positive associations in the gold standard (de-duplicated) | 1,642 |
| Negative associations in the gold standard (de-duplicated) | 23,737 |

## 3.6 Gold Standard Development

The gold standard required for this study was developed by senior year medical students, who were presented with the pairs of unique entities from the data set and were asked to indicate whether a pair has an association or not. A physician (an MD) gave guidelines and examples to the students for the manual assessment. The students were instructed to identify any direct relationship between a pair. For example, the instructions allowed relationships such as a medication may treat or prevent a disease, or may cause a disease as a side effect. From the initial trials, it became obvious that the association is mostly independent of a patient and therefore any duplicate pairs in the aggregated data were eliminated and the students were asked to assess the relationship independent of the patient record from which the pair was drawn. Each association was assessed by two students and any conflicts were resolved by the MD. The gold standard was later vetted once the automated methods were run on this data, and corrections, if any, were made to the gold standard. The final gold standard contained 25,379 unique disease-medication pairs, including 1,642 positive instances and 23,737 negative instances.

## 3.7 Experiments, Accuracy Metrics, and Analysis

We used a 10 x 10 cross validation to conduct accuracy analysis. The model was always trained using an equal number of positive and negative pairs, but the model is tested on the imbalanced set. The results are reported for the aggregate of all 10-fold cross validation iterations.

In the experiments, we tested the performance of four methods: the baseline method described earlier, DRE, AD, and a stacked ensemble of DRE and AD. Each experiment involved obtaining the association scores for each of the four methods for the entity pairs in the test set, using a threshold to determine if the association is positive or negative as per the method. Association scores range from 0 to 1, and a scored association is positive if the predicted score is greater than or equal to the threshold, and negative otherwise. Using the gold standard, we then computed true positives, false positives, false negatives, and true negatives, from which we computed standard precision (P), recall (R), and F score (F1) for positive and negative associations. Note that if a method has no coverage for an entity pair, then a zero score is returned by the method and hence results in a negative association.

F1 scores for positive associations were determined and plotted at threshold values from 0.1 to 0.9, in intervals of 0.1. We also plotted the precision-recall curves and compared areas under the curves. The threshold values that optimize F1 for each method were obtained and used in the final performance comparison of the methods.

As one may recall from Section 3.5, the data contains multiple instances of some problem-medication pairs since the entity pairs are extracted from 100 patient records. For example, several patients may have been diagnosed with Diabetes and many of them may be prescribed Metformin as a treatment, in which case, the entity pair Diabetes-Metformin may occur several times in the data. Using the frequency of such occurrences in the original data as a weighting function, we determined weighted accuracy of the methods.

## 4    Results and Discussion

### 4.1    Coverage

Table 2 shows the coverage for the baseline and AD methods. Coverage is the percentage of the entity pairs in the data for which the underlying methods and sources entail a positive or negative association. DRE, as it uses UMLS CUIs and types as some of its features, always returns non-empty values for the features. On the other hand, AD and the baseline method end up with all empty features, for at least some entity pairs. For example, if a medication is never prescribed for a disease, the patient records would never have any data for it.

For the positive associations, AD has a very high coverage (88%), but the baseline method has only 43% coverage. This reflects in the baseline method's poor accuracy in predicting positive associations. For the negative associations, both have poor coverage but it does not matter as much since the default score of 0.0 would end up being a correct prediction for negative associations.

*Table 2. Coverage for the baseline and AD methods*

| Method | Coverage | |
|---|---|---|
| | Positive Associations | Negative Associations |
| Baseline | 43% | 12% |
| AD | 88% | 41% |

### 4.2    Accuracy, Thresholds, and P-R Curves

First, consider the positive association at the optimum threshold values, as shown in Table 3. For the unique (i.e. unweighted) pairs, DRE performed slightly better than AD with an F1 score of 0.60 compared to an F1 score of 0.56 for AD. Both methods performed significantly better than the baseline, which achieved an F1 score of only 0.35. The stacked method of DRE and AD performed better than the individual

*Table 3. Accuracy analysis of the methods in predicting the positive associations*

| Method | Optimum Threshold | Unweighted | | | Weighted | | |
|---|---|---|---|---|---|---|---|
| | | Precision | Recall | F1 Score | Precision | Recall | F1 Score |
| Baseline | 0.26 | 0.28 | 0.48 | 0.35 | 0.58 | 0.53 | 0.55 |
| DRE | 0.20 | 0.56 | 0.66 | 0.60 | 0.68 | 0.62 | 0.67 |
| AD | 0.26 | 0.62 | 0.52 | 0.56 | 0.72 | 0.63 | 0.67 |
| DRE + AD | 0.29 | 0.57 | 0.71 | 0.69 | 0.77 | 0.73 | 0.75 |

methods, achieving an F1 score of 0.69. Among the individual methods, AD had higher precision (0.62) and DRE had higher recall (0.66).

When weighted entity pairs are considered, which represent the frequency of occurrence of the entity pairs in the patient records we used for this study, the performance pattern of the methods generally remained unchanged. DRE and AD performed significantly better than the baseline; each achieving an F1 score of 0.67. The stacked ensemble achieved the highest F1 score of 0.75. All methods achieved higher precision than recall. We handled the imbalanced nature of the dataset by learning a threshold value that optimizes the F1 score for positive associations, which is considered one of the effective ways to deal with imbalanced datasets [21]. The optimum thresholds are shown in Table 3 for the various methods, and Figure 4 shows how the F1 score varies with the threshold for the methods.

*Table 4. An ablation study of accuracy for unique (unweighted) positive pairs*

| Method and Feature Ablation | Precision | Recall | F1 Score |
|---|---|---|---|
| AD – without drug class features | 0.60 | 0.46 | 0.52 |
| AD – with only drug class features | 0.68 | 0.42 | 0.52 |
| DRE – with only LSA features | 0.50 | 0.25 | 0.33 |
| DRE – with only UMLS features | 0.21 | 0.44 | 0.28 |
| DRE – with only DS features | 0.60 | 0.51 | 0.55 |
| DRE – with only DS features, but without argument expansion | 0.52 | 0.46 | 0.42 |

An ablation study of feature groups for the unique (i.e. unweighted) positive pairs is shown in Table 4. We removed (ablated) a selected group of logically related features and determined the accuracy which would show the importance of the feature group to the model. For AD, using either drug-class or individual drugs alone in calculating feature scores achieved the same accuracy, but when used together they improved the overall accuracy. For DRE, the distributional semantics (DS) features with argument expansion achieved accuracy close to the best DRE accuracy, and argument expansion by itself contributed significantly to the accuracy of the DS features. However, the UMLS features alone achieved the lowest F1 score, although they were useful in improving recall.

In predicting negative associations, all methods achieved very high performance for both unweighted and weighted cases. This result merely reflects the fact that all methods return 0.0 when the underlying sources provide no information for the arguments, which happens to correctly predict a negative association. It is good to see that the default in these cases does no harm because a knowledge resource or method needs to handle all scenarios well in clinical applications.

Precision-recall curves for the methods are shown in Figure 5. As precision improves, recall reduces, at different rates for the different algorithms, which reflects in the area under the curve (AUC) metrics.

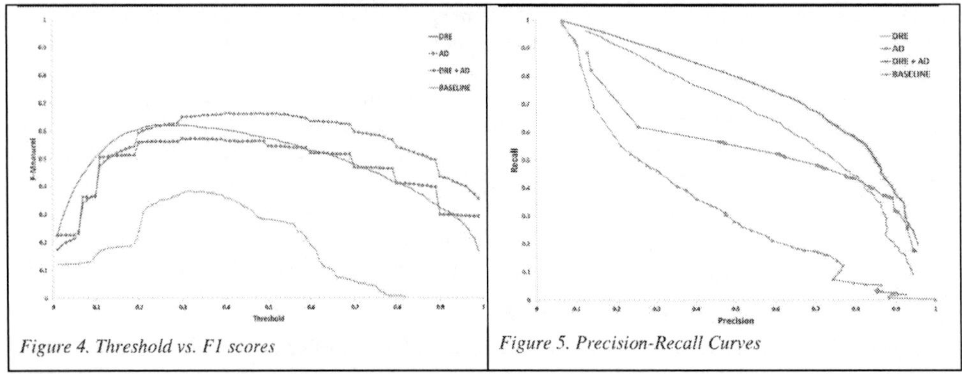

<table>
<tr><td>Figure 4. Threshold vs. F1 scores</td><td>Figure 5. Precision-Recall Curves</td></tr>
</table>

The ability of the ensemble method to improve precision while not losing recall as rapidly is helped by AD and resulted in the overall high F1 score.

### 4.3    Discussion

While each of the two methods we evaluated achieved a reasonable level of accuracy, the ensemble method achieved the best performance. The two sources and methods complement each other, forming a more effective method for association scoring. DRE relies on carefully written medical text and manually curated knowledge resources, while AD relies on statistical measures of patient care data. The intrinsic nature of the resources used by the methods reflects in the performance of the methods on certain types of entities. For example, DRE is better at coverage on rare diseases, such as scoring the pair *hypophosphatemic rickets* and *calcitriol*, and for over the counter (OTC) medications. AD performs better than DRE when not all medications within a class are equally used to treat a problem. For example, for the pair *migraine and headache syndromes* and *Inderal la*, DRE scored 0.007 whereas AD scored 0.773, which is a true positive.

### 5    Conclusion

To reduce the cognitive load on physicians in using large amounts of data in the modern EHR systems, it is necessary to imbue clinical applications with fundamental medical knowledge, such as the relationships between diseases and medications. This paper presented two new methods that used different ways of extracting features (distributional semantics and data mining) and two different content sources (large medical context and patient records) for the task. We compared the accuracy of these distinctly different approaches and their ensemble with a baseline method published previously. The results showed that an ensemble provides an accurate relation scoring system because of individual methods leveraging different content sources and feature extraction. It can be used as an on-demand scoring system, or as a method to generate association scores for a large set of entities a priori for later use in clinical decision support applications. The methods introduced here are promising, and can be expanded in the future to score specific relations such as "treats" and "prevents", and to score relations between other types of clinical data.

# References


[1]  R. Wachter, The Digital Doctor, McGraw-Hill, 2014.

[2]  T. D. Shanafelt, L. N. Dyrbye, C. Sinskye, O. Hasan, D. Satele, J. Sloan and C. P. West, "Relationship Between Clerical Burden and Characteristics of the Electronic Environment With Physician Burnout and Professional Satisfaction," *Mayo Clinic Proceedings,* vol. 91, no. 7, pp. 836-848, 2016.

[3]  US National Library of Medicine, "UMLS Reference Manual," National Library of Medicine (US), September 2009. [Online]. Available: http://www.ncbi.nlm.nih.gov/books/NBK9675/. [Accessed 15 04 2014].

[4]  J. Hirschberg and C. D. Manning, "Advances in natural language processing," *Science,* vol. 349, no. 6245, pp. 261-266, 2015.

[5]  O. Uzuner, B. R. South, S. Shen and D. L. Scott, "2010 i2b2/VA challenge on concepts, assertions, and relations in clinical text," *Journal of the American Medical Informatics Association,* vol. 18, no. 5, pp. 552-556, 2011.

[6]  C. Wang and J. Fan, "Medical Relation Extraction with Manifold Models," in *The 52nd Annual Meeting of the Association for Computational Linguistics (ACL 2014),* 2014.

[7]  W.-Q. Wei, R. M. Cronin, H. Xu, T. A. Lasko, L. Bastarache and J. C. Denny, "Development and evaluation of an ensemble resource linking medications to their indications," *Journal of the American Medical Informatics Association : JAMIA,* vol. 20, no. 5, pp. 954-961, 2013.

[8]  A. Wright, E. S. Chen and F. L. Maloney, "An automated technique for identifying associations between medications,," *Journal of Biomedical Informatics,* vol. 43, pp. 891-901, 2010.

[9]  F. Severac, E. A. Sauleau, N. Meyer, H. Lefevre, G. Nisand and N. Jay, "Non-redundant association rules between diseases and medications: an automated method for knowledge base construction," *BMC Medical Informatics and Decision Making,* vol. 15, no. 29, 2015.

[10] H. Salmasian, T. H. Tran, H. S. Chase and C. Friedman, "Medication-indication knowledge bases: a systematic review and critical appraisal," *Journal of the American Medical Informatics Association,* vol. 22, no. 6, pp. 1261-1270, 2015.

[11] C. Biemann and M. Riedl, "Text: now in 2D! A framework for lexical expansion with contextual similarity," *Journal of Linguistic Modeling,* vol. 1, no. 1, pp. 55-95, 2013.

[12] A. Gliozzo, "Beyond Jeopardy! Adapting Watson to New Domains Using Distributional Semantics," 2013. [Online]. Available: https://www.icsi.berkeley.edu/icsi/sites/default/files/events/talk_20121109_gliozzo.pdf. [Accessed 18 04 2014].

[13] D. Ferrucci, A. Levas, S. Bagchi, D. Gondek and R. T. Mueller, "Watson: Beyond Jeopardy!," *Artificial Intelligence,* pp. 93-105, 2013.

[14] A. M. Gliozzo, M. Pennacchiotti and P. Pantel, "The domain restriction hypothesis: Relating term similarity and semantic consistency," in *Proceedings of HLT-NAACL,* Rochester, NY, 2007.

[15] S. Deerwester, D. T. Susan, G. W. Furnas, T. K. Landauer and R. Harshman, "Indexing by Latent Semantic Analysis," *Journal of the American Society for Information Science,* vol. 41, no. 6, pp. 391-407, September 1990.

[16] S. Simmons and Z. Estes, "Using latent semantic analysis to estimate similarity," Hillsdale, NJ, 2006.

[17] L. Breiman, "Random Forests," *Machine Learning,* vol. 45, no. 1, pp. 5-32, 2001.

[18] R.-E. Fan, K.-W. Chang, C.-J. Hsieh, X.-R. Wang and C.-J. Lin, "Liblinear: A library for large linear classification.," *Journal of Machine Learning Research,* vol. 9, p. 1871–1874, 2008.

[19] M. Devarakonda and C.-H. Tsou, "Automated Problem List Generation from Electronic Medical Records in IBM Watson," in *Proceedings of the Twenty-Seventh Conference on Innovative Applications of Artificial Intelligence,* Autin, TX, 2015.

[20] M. V. Devarakonda and N. Mehta, "Cognitive Computing for Electronic Medical Records," in *Healthcare Information Management Systems, 4th Edition,* A. C. Weaver, J. M. Ball, R. G. Kim and M. J. Kiel, Eds., Springer International, 2015.

[21] W. Klement, S. Wilk, W. Michaowski and S. Matwin, "Classifying Severely Imbalanced Data," in *Advances in Artificial Intelligence, Proc. of 24th Canadian Conf on Artificial Intelligence,* St. John's, Springer Berlin Heidelberg, 2011, pp. 258-264.

# Author Name Disambiguation in MEDLINE Based on Journal Descriptors and Semantic Types


**Dina Vishnyakova**
Roche Pharmaceutical Research and Early Development, pRED Informatics, Roche Innovation Center, Basel, Switzerland and Swiss Institute of Bioinformatics, Switzerland
Dina.Vishnyakova@roche.com

**Raul Rodriguez-Esteban**
Roche Pharmaceutical Research and Early Development, pRED Informatics, Roche Innovation Center, Basel, Switzerland
raul.rodriguez-esteban@roche.com

**Khan Ozol**
Novartis Pharmaceuticals, Basel, Switzerland
khan.ozol@novartis.com

**Fabio Rinaldi**
Institute of Computational Linguistics University of Zurich and Swiss Institute of Bioinformatics, Switzerland
fabio.rinaldi@uzh.ch


## Abstract


Author name disambiguation (AND) in publication and citation resources is a well-known problem. Often, information about email address and other details in the affiliation is missing. In cases where such information is not available, identifying the authorship of publications becomes very challenging. Consequently, there have been attempts to resolve such cases by utilizing external resources as references. However, such external resources are heterogeneous and are not always reliable regarding the correctness of information. To solve the AND task, especially when information about an author is not complete we suggest the use of new features such as journal descriptors (JD) and semantic types (ST). The evaluation of different feature models shows that their inclusion has an impact equivalent to that of other important features such as email address. Using such features we show that our system outperforms the state of the art.


## 1    Introduction

A frequent task for researchers is searching for relevant publications or citations. These resources are often queried by the name of an author. According to Dogan et al. (2009) queries based on Author Name are most frequent in PubMed and make approximately 36% of all queries. However, author names can be highly ambiguous, which complicates any author search and posterior analysis. Although some online literature resources partially disambiguate author names-for example, PubMed started to rank authors according to the likelihood that they are relevant to a user author name query since 2012 (Liu et al., 2014) - this is not yet an established practice. Moreover, when querying for particular topics or subjects in PubMed it is very challenging for a user to figure out the key authors relevant to the query and PubMed does not offer any aid in that respect.




*Proceedings of the Fifth Workshop on Building and Evaluating Resources for Biomedical Text Mining (BioTxtM 2016),*

Several articles regarding Author Name Disambiguation (AND) solutions in MEDLINE have been published, e.g. (Smalheiser and Torvik, 2009; Torvik et al., 2005; M. Song et al., 2015; Liu et al., 2014; Li et al., 2012; Warner, 2010; Treeratpituk and Giles, 2009). However, the AND problem is not yet satisfactorily solved. Alternatively, unique identifiers for authors such as those from *Scopus* or *ORCID* (Haak et al., 2012) have been created in order to disambiguate names in publications. However, a unique author identifier is not a requisite for publishing (Smalheiser and Torvik, 2009). Moreover, some existing unique identifiers assigned to authors by citation or abstract databases such as *Scopus* or *arXiv Author ID* (Warner, 2010) are based on an automatic information extraction mechanism and often are not validated by the authors themselves, and therefore can contain errors.

The majority of the methods described in (Torvik et al., 2005; M. Song et al., 2015; Liu et al., 2014; Li et al., 2012; Warner, 2010; Treeratpituk and Giles, 2009) base their disambiguation methods on author personal data from MEDLINE records such as name, affiliation, co-authorship and e-mail address (Torvik et al., 2005; M. Song et al., 2015; Liu et al., 2014; Cota et al., 2010). While information regarding an author's last name and first name is an essential part of a scientific article, information regarding the author's affiliation is not always provided by MEDLINE. As an example, (Liu et al., 2014) mentions that information about affiliation was available only in 53% of the publications they considered. Beyond personal data, information such as MeSH terms and keywords has also been used for disambiguation. According to (Liu et al., 2014), the availability of MeSH terms in MEDLINE is ~ 91%, which is larger, in the sense of publication coverage, than the availability of affiliation information.

Commonly, disambiguation methods estimate author publications within the same "equivalence set," where each set is defined by all the authors that share the same last name and first initial. This means that author publications need to be grouped first by last name and first initial (Torvik and Smalheiser, 2009; Liu et al., 2014; M. Song et al., 2015; Cota et al., 2010). We will refer to such equivalence sets as "namespaces." Thus, identifying the namespaces is the initial and most important step for AND. Thereafter, the methods to disambiguate authorship can vary depending on the features used, which are selected to calculate inter-publication similarity. Evidently, the process of assigning an author's publication to a namespace may affect the overall results of the disambiguation.

Usually authors tend to publish their work in specific journals, conferences, workshops, etc. depending on the topics of the journals and the research domain of the author. However, in the era of translational research it becomes problematic to strictly define which topics belong to which author. This can be done, for example, through the analysis of the keywords or MeSH terms used in the author's publications or by creating author-journal similarity profiles (Torvik et al., 2005; Y. Song et al., 2007). However, when the paper has several authors, the identification of the main topics of interest of each author/co-author becomes challenging. Moreover, publications written by specialists from different domains collaborating on a common project may include key terms from different fields/domains. We propose, instead, to use journal descriptors (JDs) to aid in AND instead of keywords mentioned in the publication. The JDs add more detail by describing the different specialties associated to the articles. They can identify not only the main domain of an article but also secondary ones.

## 2 Methods

In this section we describe the features we used and their provenance for creating "author profiles." By an author profile we mean an array with the following values associated to a particular publication: 1) Last name, 2) First name, 3) Initials, 4) Publication ID (PMID), 5) Year of publication, 6) Language of the publication, 7) Title of the publication, 8) Abstract, 9) MeSH terms,  and 10) Affiliation.

### 2.1 MEDLINE information

Initially, all information available in MEDLINE regarding the author of each publication was extracted. This information includes the following: 1) last name, 2) full first name, 3) initials, 4) affiliation, 5) co-authors, 6) order of the author in the author list, 7) language of the publication, 8) MeSH terms, 9) abstract and 10) title.

Information regarding organization, city, country as well as email address were extracted from the author's affiliation. To extract the email address from the affiliation, a regular expression was used. In order to extract the organization name, the Stanford named-entity recognizer (NER) based on the 7-class model (Finkel et al., 2005) was used. This model has been trained on the MUC6 (https://catalog.ldc.upenn.edu/LDC2003T13) and MUC7 training data (https://catalog.ldc.upenn.edu/LDC2001T02). The model recognizes location, organization, person, date, money, percent and time information in text. The choice of this NER algorithm can be explained by its better performance compared to OpenNLP (Dlugolinský et al., 2013). Since affiliation information is usually represented as a short text string it was important to choose the NER model which could recognize entities with a better accuracy in such strings. A preliminary test of 3-, 4- and 7-class models for organization and location entities showed that the 7-class model outperformed other models. Then, each recognized organization was classified according to its type: 1) University, 2) School, 3) Ministry, 4) Institute, 5) Commercial Company, 6) Centre and 7) Hospital, as well as according to the type of the main descriptor of the organization. The following types of descriptors were used: 1) Chemistry, 2) Biology, 3) Psychology, 4) Health, 5) Medicine/Medical, 6) Pediatric, 7) Surgery, 8) Genetic, 9) Infection, 10) Agriculture, 11) Entomology, 12) Biotechnology, 13) Neurology, 14) Psychology, 15) Pharmacology, 16) Toxicology, 17) Nutrition and 18) Dentistry. An organization belongs to one or another type of descriptor if there is a match between the name of the organization and the name of one of the above descriptors. The organization types and descriptors represent qualitative information and were manually selected based on their observed frequency in the affiliation field. They were mapped to a numeric representation, e.g. types from 1 to 7 and descriptors from 1 to 18.

The Stanford NER was not used for country and city recognition, since the process to identify those entities in such short texts was error-prone. Instead, a dictionary-based method was used. The names of countries and cities were extracted from http://www.geonames.org/. This resource provides a list of city names in different languages. Each city name in the list is mapped to the country name. Thus, we could identify a country associated to the affiliation even in cases where the country name was missing in the affiliation.

### 2.2 Journal Descriptors and Semantic Types

Frequently, the first author in collaborative publications is the principal contributor in the research work. Other authors can present expertise from different domains. Therefore it is insufficient to measure the similarity of text taken from titles and abstracts for the purpose of AND. To complement this, we used additional descriptors to further define the content of the work. For this purpose a JDI (Journal Descriptor Indexing) tool (Humphrey et al., 2006) developed at the National Library of Medicine (NLM) was used. This tool returns a ranked list of journal descriptors (JD) or UMLS Semantic Types (ST) corresponding to biomedical descriptors as an output to a given text. Ranked items in the output have a score in a range from 0 to 1. There are overall 122 JDs and 135 STs.

| Rank | Score | | Journal Descriptor | | Descriptor | |
|---|---|---|---|---|---|---|
| | PMID 24782557 | PMID 24481031 | PMID 24782557 | PMID 24481031 | PMID 24782557 | PMID 24481031 |
| 1 | 0.0178087 | 0.1916517 | JD148 | JD148 | Pulmonary Medicine | Pulmonary Medicine |
| 2 | 0.0140019 | 0.0257541 | JD100 | JD129 | Radiology | Pathology |
| 3 | 0.0113613 | 0.0206357 | JD023 | JD144 | Communicable Diseases | Neoplasms |

Table 1. Journal Descriptors as output of the JDI tool.

Originally this tool was developed for text categorization purposes with the goal of improving information retrieval. For the AND task an abstract, a title and MeSH terms of articles were provided as an input to JDI. As an example the title, abstract and MeSH terms of the articles with PubMed ID 24481031 and 24782557 were used as input to the JDI tool and the output based on documents counts (Humphrey et al., 2006) is represented either as journal descriptors or semantic types in Tables 1 and 2. In this case the articles were published in the journals "American College of Chest Physicians" and "Respiratory Care," respectively. Both publications share only one MeSH term – "Humans", which is too common and appears in most publications. As it can be seen, the JDs and STs derived from these publications are more descriptive.

Preliminary experiments showed that in most cases the top 3 JDIs have an assigned score much higher than the other JDIs returned. Thus, only the top 3 results were used as an additional feature to describe the domain of a publication.

| Rank | Score | | UMLS Type | | Semantic Type | |
|---|---|---|---|---|---|---|
| | PMID 24782557 | PMID 24481031 | PMID 24782557 | PMID 24481031 | PMID 24782557 | PMID 24481031 |
| 1 | 0.5323717 | 0.6212719 | T046 | T203 | Pathologic Function | Drug Delivery Device |
| 2 | 0.5264287 | 0.4946694 | T185 | T082 | Classification | Spatial Concept |
| 3 | 0.5214509 | 0.4894958 | T169 | T046 | Functional Concept | Pathologic Function |

Table 2. Semantic Types as output of the JDI tool.

## 2.3 Supervised classifiers

We transform the AND problem to a binary classification task in which a classifier predicts whether the authors of two different publications are the same person. For this purpose, four well-known supervised algorithms (SVM, Random Forest, k-NN and J48) were used to do the classification as well as to evaluate the impact of the features based on Journal Descriptor and Semantic Type to the overall disambiguation performance. These classification algorithms are frequently used in data mining and text-mining tasks (Fernández-Delgado et al., 2014). They have also been used by (Han et al., 2004; Treeratpituk and Giles, 2009; M. Song et al., 2015) for the AND task. The J48 algorithm is a java implementation of the C4.5 algorithm (Quinlan, 2014). All features were normalized according to range of 0 to 1.

### 2.3.1 *Similarity pairs*

An author profile is represented as an array of values extracted from MEDLINE (name, affiliation, year of publication, etc.), journal descriptors and semantic types. Author profiles are grouped by namespaces. For each namespace, the profiles are compared in a pairwise manner, so that each pair of profiles is represented as a vector of similarity scores between the two profiles. Table 3 shows the process used to transform the discrete values of two profiles into a numeric similarity vector. A Jaro-Winkler algorithm was used to calculate similarity scores for first names of authors. The choice of this algorithm can be explained by its good performance on short strings (M. Song et al., 2015). We chose the SoftTFIDF Jaro-Winkler method to calculate a similarity score for the organizations due to its better performance on longer strings and the fact that it is a less time-consuming algorithm (Cohen et al., 2003). Finally, organization type and journal descriptor were mapped to numeric values and the difference between them was used in the similarity vector.

| Profile Values | Similarity Vector Features |
| --- | --- |
| Full First Name | Jaro-Winkler score (Full_First_Name$_a$, Full_First_Name$_b$) |
| Initials | Boolean score (Initials$_a$, Inititals$_b$) |
| Co-Authors | # of shared co-author names |
| MeSH terms | # of shared MeSH terms |
| JDI (3 entities) | # of shared descriptors or semantic types |
| City | "1" (City$_a$ = City$_b$), "0" (City$_a \neq$ City$_b$) |
| Country | "1" (Country$_a$ = Country$_b$), "0" (Country$_a \neq$ Country$_b$) |
| Language | "1" (Lang$_a$ = Lang$_b$), "0" (Lang$_a \neq$ Lang$_b$) |
| Year | \|Year$_a$ – Year$_b$\| |
| Organisation | SoftTFIDF Jaro-Winkler score (Organisation$_a$, Organisation$_b$) |
| Email | "1" (email$_A$ = email$_B$), "0" (email$_A \neq$ email$_B$) |
| Type and Descriptor of Organisation | diff (Type$_a$ Descriptor$_a$, Type$_b$, Descriptor$_b$) |

Table 3. Similarity vector used to compare the profiles of two authors *a* and *b*.

## 3    Data

To evaluate the classifiers, a curated corpus for author name disambiguation was used (M. Song et al., 2015). The dataset contains 2,875 publications authored by 385 first authors with 431 author name variants. In less than half of the publications information about emails is present. Furthermore, the majority of the names are of Western origin. Each author in the list has a unique ID assigned by the dataset providers. To date, this is the only known dataset for AND in MEDLINE which is manually curated.

Since the original dataset only consist of author names, PubMed IDs and author IDs, it was necessary to extract all additional relevant information from the MEDLINE corpus. Our final dataset is based on the 2014 MEDLINE/PubMed Baseline Database Distribution. Because the authors considered are only first authors, affiliations are available for the majority of them.

There are articles in 5 different languages in the dataset (denoting the main language of the article's full text, not of the abstract): English, Japanese, Chinese, German and French. The earliest publications are dated from 1967 and the most recent from 2013.

After transformation of pairs of author profiles to similarity vectors, less than a quarter of them belonged to the positive class, i.e. they correspond to the same authors.

## 4    Results

In this section we present results for each classifier using 10-fold cross-validation. Further, we provide the results of the classifiers from (M. Song et al., 2015) for comparison. Then, we show evaluation scores for the features used in the disambiguation process to rank them according to their contribution.

### 4.1    Classifier performance

Tables 4-7 show the results obtained by the classifiers. These results are based on three models used to train the classifiers with the following features: (1) Medline features and journal descriptors (MF+JD) obtained with the JDI tool, (2) MF and semantic types (MF+ST) and (3) MF, JD and ST (MF+JD+ST). Additionally we provide the results of the Named Entity Recognizer-based model (NER-based model) and Baseline model described in (Song, Kim et al. 2015). Song's NER model is based on MEDLINE features such as author name, co-authors, affiliation and keywords extracted from article title and journal title. Additionally, Song's NER model relied on the output of the Stanford NER algorithm, which identified organizations, locations and emails in the affiliation text.

Thus, detected entities were transformed into features. Song's Baseline model (M. Song et al., 2015) is based on first author name, article title, and publication venue.

| Metrics | MF+JD | MF+ST | MF+JD+ST | NER-Based | Baseline |
| --- | --- | --- | --- | --- | --- |
| Precision | 0.986 | 0.975 | 0.987 | 0.9776 | 0.8348 |
| Recall | 0.992 | 0.961 | 0.994 | 0.9545 | 0.8501 |
| F-Measure | 0.989 | 0.9675 | 0.990 | 0.9657 | 0.8423 |

Table 4. Results of the J48 classifier.

| Metrics | MF+JD | MF+ST | MF+JD+ST | NER-Based | Baseline |
| --- | --- | --- | --- | --- | --- |
| Precision | 0.9785 | 0.9785 | 0.991 | 0.9884 | 0.8349 |
| Recall | 0.9685 | 0.9725 | 0.996 | 0.9634 | 0.8499 |
| F-Measure | 0.973 | 0.978 | 0.993 | 0.9756 | 0.8322 |

Table 5. Results of the Random Forest classifier.

| Metrics | MF+JD | MF+ST | MF+JD+ST | NER-Based | Baseline |
| --- | --- | --- | --- | --- | --- |
| Precision | 0.985 | 0.956 | 0.987 | 0.9723 | 0.8253 |
| Recall | 0.988 | 0.951 | 0.977 | 0.9595 | 0.8412 |
| F-Measure | 0.986 | 0.9535 | 0.982 | 0.9656 | 0.8330 |

Table 6. Results of the k-NN classifier.

The results achieved on the MF+JD+ST model show a recall which is slightly higher than the precision. In the results of the NER-model the precision has a tiny prevalence over the recall. In Table 7, the precision achieved on models MF+ST and MF+JD+ST is a little greater than the recall, though it is the opposite for the model MF+JD.

| Metrics | MF+JD | MF+ST | MF+JD+ST | NER-Based | Baseline |
| --- | --- | --- | --- | --- | --- |
| Precision | 0.964 | 0.949 | 0.9695 | 0.9541 | 0.8353 |
| Recall | 0.991 | 0.894 | 0.905 | 0.8385 | 0.8478 |
| F-Measure | 0.977 | 0.9185 | 0.9335 | 0.8870 | 0.8414 |

Table 7. Results of the SVM with RBF kernel.

## 4.2 Features Contribution

The information gain feature provided by WEKA was used in order to compute the value of a feature attribute by measuring the information gain with respect to the class. The ranked list of features and their impact according to the information gain is shown in Tables 8 and 9.

The value of the information gain of semantic types is less than that of journal descriptors; see Table 8 and Table 9. In both tables the value of the co-author, year and MeSH-terms features are almost equivalent.

| Profile Features | Rank | Value |
| --- | --- | --- |
| Full First Name | 1 | 0.310439 |
| Organisation | 2 | 0.292023 |
| Email | 3 | 0.214672 |
| JDIs | 4 | 0.202067 |
| Type and Descriptor of Organization | 5 | 0.183693 |
| Co-Authors | 6 | 0.152621 |
| City | 7 | 0.022 |
| Initials | 8 | 0.01097 |
| Year | 9 | 0.010227 |
| Language | 10 | 0.000584 |
| Country | 11 | 0.000532 |
| MeSH Terms | 12 | 0.0 |

Table 8. Ranked list of the information gain of features with respect to the class in the MF+JD model.

| Profile Features | Rank | Value |
| --- | --- | --- |
| Organization | 1 | 0.35203 |
| Full First Name | 2 | 0.287596 |
| Email | 3 | 0.255492 |
| Type and Descriptor of Organization | 4 | 0.20955 |
| Co-Authors | 5 | 0.154428 |
| Semantic Types | 6 | 0.119648 |
| City | 7 | 0.034587 |
| Year | 8 | 0.010847 |
| Initials | 9 | 0.009418 |
| Country | 10 | 0.006007 |
| Language | 11 | 0.000532 |
| MeSH Terms | 12 | 0.0 |

Table 9. Ranked list of the information gain of features with respect to the class in the MF+ST model.

## 5 Discussions

The evaluation was performed on the dataset in three different ways: (1) MF+JD, (2) MF+ST and (3) MF+JD+ST. Moreover, we have compared the results to the ones obtained by (M. Song et al., 2015) on the Baseline and NER-based models. Our evaluation results show that the classifiers J48 and Random Forest performed better than the rest. Random Forest provided slightly better results, but in terms of time it was slower than J48. This can be explained by the number of training trees used in the process. The comparison of overall results to Song's NER-Model shows that a significant difference in scores is achieved by our SVM algorithm. However, compared to other classifiers, SVM is less efficient for the AND task and most time-consuming. These results could be explained by the low dimensionality of our data, since SVM performs better on highly dimensional data

The results show also that the MF+JD+ST model outperformed the other models using features based on the topics or descriptors rather than on the keywords or MeSH terms. Nevertheless, the results of the MF+ST model differ from those of the MF+JD. Despite the assumption that the semantic description of the publication would better represent the content, the semantic types and the model MF+ST did not add significant improvement over the MF+JD results. Possible reasons for these results include the fact that the results of the model MF+JD are already sufficiently good, and also that semantic types offer a better characterization of papers than keywords. Surprisingly, the MeSH terms, according to the feature estimation results, showed no impact on the disambiguation model. The information gain of feature attributes also shows that even though the ST-based feature has a higher impact compared to year, language and location, it only brings slight improvements to the classification results based on the results from the MF+JD+ST model.

The assumption that the location of the author can help disambiguate two profiles was not confirmed. It is not rare when authors change their affiliation and consequently their location. However, in cases where the location of two profiles is identical it suggests that these profiles share the same authorship. An email address, nonetheless, is more significant than a location. The evaluation of features surprisingly showed that journal descriptors and topics are as useful as email addresses for the disambiguation process. Considering that information about the email address of the author is often missing, then the feature based on the journal descriptor and topics can still be used to disambiguate an ambiguous author name.

## 6    Conclusion

In this paper we have introduced new disambiguation features such as journal descriptors and semantic types, which were not previously used for Author Name Disambiguation. Classification was done with the four most used classifiers for the AND task. The achieved results were compared to state-of-the-art results and it was shown that journal descriptors are as helpful in the disambiguation process as email addresses.   Regarding the unexpected value of the MeSH Terms for the classification, the impact of the semantic types to the model can be explained by their nature. Unlike MeSH terms, they are automatically generated for each articles and their granularity is greater.

It is worth mentioning that the results of the evaluation are achieved on the so-called gold standard dataset provided by (M. Song et al., 2015). To date, this is the only dataset which is manually verified. One of the disadvantages of this dataset is that it consists of only first authors of publications. Consequently, the results may be less competitive if datasets consists not only of first authors but also of co-authors.   Indeed, in MEDLINE the information about affiliation of co-authors is frequently missing. Moreover, the dataset is biased towards Western types of last names, e.g. Smith, Cohen, Taylor. However, the statistics of most frequent author names in MEDLINE show that they are of Asian origin, for example Wang, Zhang, etc. If we consider that, in the 1990 edition of the Guinness Book of World Records, Zhang was the most common last name in the world, then the disambiguation of certain Asian author names seems extremely challenging. Thus, the classifier models trained on the gold-standard dataset are not necessarily applicable to the AND task for the entire MEDLINE, where non-first authors have missing affiliation and most frequent names are ethnicity-sensitive to the name-matching process (Treeratpituk and Giles, 2012; Jimenez-Contreras et al., 2002; Louppe et al., 2015; Kim and Cho, 2013).

## Reference


Cohen, W., Ravikumar, P., & Fienberg, S. (2003). *A comparison of string metrics for matching names and records.* Paper presented at the Kdd workshop on data cleaning and object consolidation.

Cota, R. G., Ferreira, A. A., Nascimento, C., Gonçalves, M. A., & Laender, A. H. (2010). An unsupervised heuristic-based hierarchical method for name disambiguation in bibliographic citations. *Journal of the American Society for Information Science and Technology, 61*(9), 1853-1870.

Dlugolinský, Š., Ciglan, M., & Laclavík, M. (2013). *Evaluation of named entity recognition tools on microposts.* Paper presented at the 2013 IEEE 17th International Conference on Intelligent Engineering Systems (INES).

Dogan, R. I., Murray, G. C., Névéol, A., & Lu, Z. (2009). Understanding PubMed® user search behavior through log analysis. *Database, 2009,* bap018.

Fernández-Delgado, M., Cernadas, E., Barro, S., & Amorim, D. (2014). Do we need hundreds of classifiers to solve real world classification problems. *J. Mach. Learn. Res, 15*(1), 3133-3181.

Finkel, J. R., Grenager, T., & Manning, C. (2005). *Incorporating non-local information into information extraction systems by gibbs sampling.* Paper presented at the Proceedings of the 43rd Annual Meeting on Association for Computational Linguistics.

Haak, L. L., Fenner, M., Paglione, L., Pentz, E., & Ratner, H. (2012). ORCID: a system to uniquely identify researchers. *Learned Publishing, 25*(4), 259-264.

Han, H., Giles, L., Zha, H., Li, C., & Tsioutsiouliklis, K. (2004). *Two supervised learning approaches for name disambiguation in author citations.* Paper presented at the Digital Libraries, 2004. Proceedings of the 2004 joint ACM/IEEE conference on.

Humphrey, S. M., Lu, C. J., Rogers, W. J., & Browne, A. C. (2006). *Journal descriptor indexing tool for categorizing text according to discipline or semantic type.* Paper presented at the AMIA Annual Symposium Proceedings.

Jimenez-Contreras, E., Ruiz-Pérez, R., & Delgado-Lopez-Cozar, E. (2002). Spanish personal name variations in national and international biomedical databases: implications for information retrieval and bibliometric studies. *Journal Medical Library Association, 90*(4).

Kim, S., & Cho, S. (2013). Characteristics of Korean personal names. *Journal of the American Society for Information Science and Technology, 64*(1), 86-95.

Li, S., Cong, G., & Miao, C. (2012). *Author name disambiguation using a new categorical distribution similarity.* Paper presented at the Joint European Conference on Machine Learning and Knowledge Discovery in Databases.

Liu, W., Islamaj Doğan, R., Kim, S., Comeau, D. C., Kim, W., Yeganova, L., et al. (2014). Author name disambiguation for PubMed. *Journal of the Association for Information Science and Technology, 65*(4), 765-781.

Louppe, G., Al-Natsheh, H., Susik, M., & Maguire, E. (2015). Ethnicity sensitive author disambiguation using semi-supervised learning. *arXiv preprint arXiv:1508.07744.*

Quinlan, J. R. (2014). *C4. 5: programs for machine learning*: Elsevier.

Smalheiser, N. R., & Torvik, V. I. (2009). Author name disambiguation. *Annual review of information science and technology, 43*(1), 1-43.

Song, M., Kim, E. H.-J., & Kim, H. J. (2015). Exploring author name disambiguation on PubMed-scale. *Journal of Informetrics, 9*(4), 924-941.

Song, Y., Huang, J., Councill, I. G., Li, J., & Giles, C. L. (2007). *Efficient topic-based unsupervised name disambiguation.* Paper presented at the Proceedings of the 7th ACM/IEEE-CS joint conference on Digital libraries.

Torvik, V. I., & Smalheiser, N. R. (2009). Author name disambiguation in MEDLINE. *ACM Transactions on Knowledge Discovery from Data (TKDD), 3*(3), 11.

Torvik, V. I., Weeber, M., Swanson, D. R., & Smalheiser, N. R. (2005). A probabilistic similarity metric for Medline records: A model for author name disambiguation. *Journal of the American Society for information science and technology, 56*(2), 140-158.

Treeratpituk, P., & Giles, C. L. (2009). *Disambiguating authors in academic publications using random forests.* Paper presented at the Proceedings of the 9th ACM/IEEE-CS joint conference on Digital libraries.

Treeratpituk, P., & Giles, C. L. (2012). *Name-Ethnicity Classification and Ethnicity-Sensitive Name Matching.* Paper presented at the AAAI.

Warner, S. (2010). Author identifiers in scholarly repositories. *arXiv preprint arXiv:1003.1345.*

# Author Index

**Association for Computational Linguistics**
209 N. Eighth Street
Stroudsburg, Pennsylvania 18360

ISBN 978-1-5108-3331-9